LN2
Cryogenic Freezing Manual
First Edition

Cornelis J.F. Elenbaas

Cryogenic Consultants

Austin, Texas, U.S.A.

 www.trafford.com

North America & international
toll-free: 1 888 232 4444 (USA & Canada)
phone: 250 383 6864 ♦ fax: 812 355 4082

WITH ALL I knew and all I tried to tell

Those wonderful places I went

I led them to graze, I taught them to spell

But why the rancour and why the tears

Perhaps, should there have been some other words

And maybe, but maybe, could I have done some good

To all my friends, do only good and you need no other words

CE

TABLE OF CONTENTS

DISCLAIMER

"CRYOGENIC FREEZING MANUAL", is made available by the author as a guide only to anybody in the process of designing a cryogenic freezing system, or is currently using such a system. Neither the author nor any representative thereof, nor any of their employees, makes any warranty, express or implied, or assumes any legal liability or responsibility for the accuracy, completeness, or usefulness of any information, apparatus, product or process disclosed, or represents that its use would not infringe privately owned right. Reference herein to any specific commercial product, process, or service by trade name, trademark, manufacturer, or otherwise, does not necessarily constitute or imply its endorsement, recommendation, or favoring by the author or any agency thereof. The views and opinions of originators expressed herein do not necessarily state or reflect those of the author or any agency thereof.

FOREWORD

MANY BOOKS HAVE been written about food freezing. In general, these books elaborate about the physical transformations that take place in food items when undergoing a temperature change. They deal with the effects on cell structures during the freezing process and the subsequent quality of the food item when thawed. No literature has been published about the technology, the mechanics, and the practical aspects of operating cryogenic food freezers. How to select such a freezer to freeze a specific food product requires an in-depth knowledge of cryogenic freezer technology. Actually, the selection of the freezers is most likely done by someone other than the person(s) using it. Often we see that the most qualified engineers ignore the basic principles of cryogenic freezing technology when it comes to selecting and operating one of the most important links in a frozen product production line, the liquid nitrogen freezer.

The purpose of this manual is to provide a source of information on experience gained from the design, installation, and operation of many previously installed freezing systems.

To take full advantage of the technology and to utilize a very powerful freezing media, liquid nitrogen, we must *first* understand how liquid nitrogen behaves in freezing operations. The liquid nitrogen (LN2) or sometimes called (LIN), is most likely delivered by over the road transports to a storage tank installed at a processing plant. From here, through a properly sized conduit, the LN2 has to be transferred to the freezer by either gravity or a pressure differential. If the tank and/or the conduit are not designed properly, the error will be irreversible. I hope that this handbook will give the plant engineer(s) enough knowledge and hence, the ability to participate in the selection and the design and most importantly the efficient operation of the cryogenic freezer.

Ironically, that by their innate simplicity, cryogenic freezers lend themselves to operator abuse.

To avoid this, I have attempted with this manual to give the operator a tool to operate the "freezing system" with optimum efficiency and safety.

It is to be used as a guide for the operators of the "freezing system" not to make the common errors seen in most operations using cryogenic freezers. The vast majority of the freezing tunnels and other cryogenic freezers observed are used improperly and subsequently are using too much LN2 and thus erroneously labeled as high cost freezing systems. This book has made its contribution to this dynamic industry if it persuades the operating engineers to adhere to the fundamentals as outlined in the following chapters and, with this knowledge, will run the freezers correctly. The main intent is to provide a simple explanation of some of the problems and experiences gained with LN2 freezing systems, and how they were resolved, and applied to future freezing system design and operations.

If more theoretical data is sought dealing with the thermodynamics of freezing foodstuffs, refer to a publication by C.P. Mallett "Frozen Food Technology" and published by Blackie Academic & Professional, London, UK.

Fig. 1   A typical air separation plant

SECTION A

SOME THEORY

1. LIQUID NITROGEN AS A REFRIGERANT

1.1 Introduction

1.2 BASIC PHYSICAL PROPERTIES OF LIQUID NITROGEN
1.2.1 *What is Nitrogen?*
1.2.2 *Properties of LN2*

1.3 THE USE OF LN2 AS A REFRIGERANT
1.3.1 *Why LN2?*
1.3.2 *Heat Absorbency of Ice*
1.3.3 *Heat Absorbency of LN2*

1.1 Introduction

As the heading implies, the scope of this chapter is to understand the mechanics of the freezer and its supporting appurtenances using liquid nitrogen (LN2) as the refrigerant. To grasp to its fullest extend how LN2 freezers are designed and should be operated, it is advisable first to understand the physical characteristics of LN2 as described in the following chapters.

> *For example: We should understand that the storage tank is the foundation on which the entire freezing system is built and in addition, that a poorly designed conduit between the tank and the freezer will render useless all the efforts made in selecting the freezer.*

An efficient freezing system has a properly designed LN2 storage tank, properly designed piping and exhaust system. A half million-dollar tunnel freezer is useless when any of the aforementioned components are poorly engineered. In the following chapters, we will explain the difference between a "Chilling" or "Freezing" tunnel and the different freezers used to accomplish these tasks.

A widely used term in the frozen food industry is "Individually Quick Frozen" products or IQF products. This term defines a product of which its individual components have been frozen without sticking to other particles of the same product. Frozen strawberries and green peas are often labeled as IQF strawberries and IQF green peas.

Liquid Nitrogen is used in different types of freezers. The freezing tunnel, the spiral freezer, the immersion or bath freezer as a stand-alone freezer and a combination of the immersion freezer with a tunnel or spiral freezer will be discussed in the following chapters. We will also see that a freezer designed to freeze a particular product is not necessarily suitable to freeze different types of products. For instance, a tunnel specifically designed to freeze hamburger patties cannot be used to freeze a battered and breaded shrimp product or a bakery product without undergoing extensive modifications.

We will describe the two different types of tunnels, the "Spray Counter flow" tunnel, and the "Convection" tunnel. Both have very different freezing characteristics, consumption rates and are used for different products.

1.2   BASIC PHYSICAL PROPERTIES OF LIQUID NITROGEN

1.2.1   *What is Nitrogen?*

Nitrogen (LN2) is an essential element for the survival of all living beings. The atmosphere at sea level is composed of approximately 78% nitrogen, 21% oxygen and about 1% of other gases. Although the human body needs nitrogen in air, high concentrations should be avoided and could be dangerous. Also, see chapter (General Safety Guidelines) on page 167.

By means of a filtration, compression and distillation process, the pure nitrogen is liquefied and stored in vacuum insulated storage tanks under a pressure of approximately 10 PSIG. A pump is used to transfer the liquefied nitrogen into tankers which than deliver and transfer the LN2 to the customer's storage tanks.

1.2.2   *Properties of LN2*

Nitrogen at atmospheric temperature and pressure is colorless, tasteless and an invisible gas; making it particularly treacherous for human beings. Cold nitrogen vapors may be visible due to the condensation of the humidity in the air.

Nitrogen has a boiling point of -320°F (-196°C) at atmospheric pressure and has its maximum energy value.

TABLE 1: PHYSICAL AND CHEMICAL PROPERTIES OF NITROGEN

PROPERTIES	CHARACTERISTICS
Molecular Weight	28
Boiling Point	**-320.4°F (-196°C)**
Heat of Vaporization	85.4 BTU/Lb.
Critical Temperature	-233°F (-147°C)
Relative Density, Gas	0.97 (Air = 1)
Relative Density, Liquid	0.80 (Water = 1)
Appearance, Gas	Invisible, odorless, tasteless
Reactivity	Inert

TABLE 2: PROPERTIES OF LN2 AT DIFFERENT TANK PRESSURES

PA	PSIG	TBP	HVAP	S.H.	CP	QTotal
14.7	0	-320.4	85.4	80.7	0.252	166.1
20	5	-315.6	83.9	80.7	0.253	163.7
25	10	-311.8	82.6	79.1	0.254	161.7
30	15	-308.6	81.5	78.6	0.255	160.1
35	***20**	**-305.7**	**80.5**	**78.1**	**0.255**	**158.6**
40	26	-303.2	79.8	77.7	0.256	157.2
50	36	-298.7	77.7	77.1	0.258	154.8
60	46	-294.8	76.2	76.6	0.260	152.8
70	56	-291.3	74.7	76.2	0.262	150.9
80	66	-288.3	73.3	75.9	0.263	149.2
90	76	-285.4	71.9	75.6	0.265	147.5
100	86	-282.8	70.7	75.4	0.267	146.1
120	105	-278.1	68.2	75.1	0.270	143.3
140	125	-274.0	65.9	74.9	0.273	140.8
160	145	-270.2	63.6	74.8	0.277	138.4

NOTE: *20 PSIG is the most common pressure used in vertical receivers.

P   = **Tank Pressure, PSIG = PSIA – 14.7**

T_{BP} = **LN2 Boiling Point Temperature (°F)**

H_{VAP} = **LN2 Heat of Vaporization (BTU/LB)**

S.H.   = **Sensible Heat (BTU/LB) from boiling point to 0°F**

C_P = **Heat Capacity (BTU/LB/°F)**

Q_{TOTAL} = **The Total Cooling power of LN2 with an exhaust temperature of 0°F.**

1.3   THE USE OF LN2 AS A REFRIGERANT

Liquid Nitrogen (LN2) has been used primarily in space research and in industrial and medical processes for some time. Since the early 60s, the food industry began extensive experimentation that eventually got us where we are today and the increased interest in cryogenics by the food industry has created an entirely new technology for the food processors.

By definition, cryogenics is the science concerned with the application of extremely low temperatures. And it has been assigned a temperature of -238°F (-150°C) which replaces conventional refrigeration.

Some of the cryogenic fluids and carbon dioxide (CO_2) and their boiling points are summarized in Table 3.

1.3.1   Why LN2?

LN2 is inert.

It is not corrosive. Only under extreme pressure and temperature can nitrogen, in the presence of certain elements, react chemically.

LN2 is safe.

It is neither explosive nor toxic.

LN2 is a rapid refrigerant.

Due to its extremely low temperature, LN2 evaporates very quickly. Thus, is capable of absorbing an unusually large amount of heat during the change of state.

LN2 is an effective preservative.

Because the gas is colorless, odorless and tasteless, LN2 as a gas is an excellent preservative of packaged foods.

TABLE 3

Normal boiling points of cryogenic fluids compared to water and CO_2 at atmospheric pressure.

(Water	+212°F)	**LN2**	**-320°**
(CO_2	-109°)	H2	-423°
O_2	-297°	Absolute Zero	-460°

1.3.2   Heat Absorbency of Ice

Coolants function according to their ability to absorb heat. Water in the form of ice is an excellent as well as inexpensive cooling agent. Its heat absorbing ability is illustrated in Figure 2.

The 193 BTU's required to transform water of 70°F to ice of 10°F also represents the amount of heat available in 1 Lb. of ice for cooling other materials.

LN2 behaves much in the same manner but operates in the cryogenic temperature range, which is far below that of ice. Naturally, its boiling point of -320°F is more easily visualized as a reference for freezing than that of ice, because this very low temperature is unfamiliar in our daily environment. The heat absorption capability of LN2 is illustrated in Fig.3

FIGURE 2

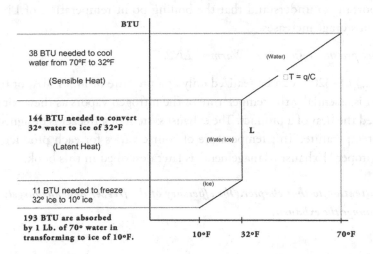

38 BTU needed to cool water from 70°F to 32°F

(Sensible Heat)

144 BTU needed to convert 32° water to ice of 32°F

(Latent Heat)

11 BTU needed to freeze 32° ice to 10° ice

193 BTU are absorbed by 1 Lb. of 70° water in transforming to ice of 10°F.

1.3.3   Heat Absorbency of LN2

FIGURE 3

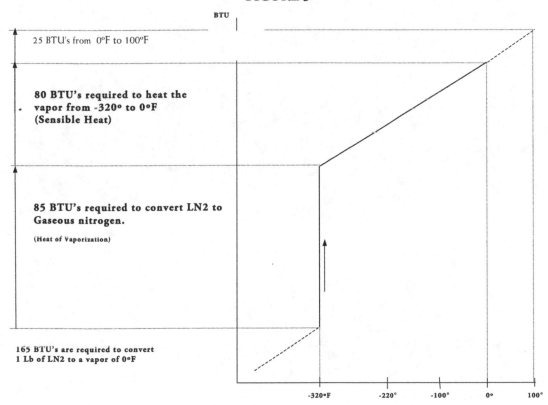

25 BTU's from 0°F to 100°F

80 BTU's required to heat the vapor from -320° to 0°F
(Sensible Heat)

85 BTU's required to convert LN2 to Gaseous nitrogen.

(Heat of Vaporization)

165 BTU's are required to convert
1 Lb of LN2 to a vapor of 0°F

In Table 2, the properties of LN2 are expressed as a function of the pressure in the storage tank. It is therefore important to understand that the boiling point temperature of LN2 increases when the pressure of storage increases.

A higher storage tank pressure results in a "Warmer" LN2.

A boiling point (T_{BP}) of -320°F can be realized only at a pressure of 14.7 PSIA, or 0 PSIG. The 0° F in the diagram is essentially the temperature of the nitrogen vapors as they exit the freezer, *after* having removed the heat of a product. The exhaust system of a freezer is designed to remove the vapors at this temperature. This temperature of course varies for each product. A detailed description of the proper "Exhaust Management" is later described in this book.

> **Pay close attention to that chapter. The efficiency of the freezer depends largely on how you manage the exhaust.**

2. SOME BASICS ABOUT THE PHYSICAL TRANSFORMATION OF FOODS DURING FREEZING

2.1 Introduction

This book has been written for a food processor freezing foodstuffs and for the one in the process of designing a system to do so. It is to his (her) benefit to have a basic understanding of the physical transformations that take place in the food item being frozen. Freezing implies to two closely linked phenomena: (a) a lowering of temperature, and (b) a change of phase from a liquid to a solid. The ultimate goal here is to reduce the temperature of the food item to lengthen its storage life. **Although from a practical point of view, our principle interest lies primarily in the phase change that transforms water to ice.**

Chilling on the other hand implies to the reduction of the product's temperature to a temperature above its freezing point. Hence, staying out of its latent heat zone.

2.2 The Freezing Process

In the practical world, when one gets about the business of freezing a product, little time is spend in understanding what is actually required to freeze a product, even less time is spend in understanding the actual thermodynamics taking place within the food item. An in-depth discussion of the theory of freezing can be found in countless books written about this phenomenon. However, one must have some basic knowledge of the freezing process and the fundamental issues relating to the "water" contend of the product and how it will influence the *rate* of freezing. Certain design criteria required of food freezing equipment are inherently related to the water or moisture contends of the food item. This will be a topic of discussion later in this book.

2.3 The Thermodynamics Of Transforming Water To Ice

When a temperature change takes place in a food item without a phase change, this change may be described with the formula:

$$\Delta T = q/C$$

Where ΔT is the product temperature change, q is the amount of heat removed, and C is the heat capacity or Specific Heat of the product.

Where a phase change take place, a further change in heat contend can take place without the accompanying change in temperature. The heat associated with this second phenomenon is known as the *latent heat*, and is expressed as L. Now lets us transform water to ice incorporating a phase change.

2.4   Pure Water To Ice

Liquid water has a heat capacity of approximately 0.998 BTU/Lb/°F (1 BTU/Lb/°F). By contrast, ice has a heat capacity of 0.5 BTU/Lb/°F. For our discussion, we assume that temperature differences are not affecting these values. At 32°F the phase change from water to ice requires 144 BTU/Lb., *the latent heat of crystallization of water.*

Example: C pure water   1 BTU/Lb/°F
    C ice   0.5 BTU/Lb/°F
    32° water to 32° ice   144 BTU/Lb.

One pound of water has a temperature of T1 = 70°F.

Freeze this water to ice of 10°F. How much heat must be removed to accomplish this?

Solution:

1.   First we must reduce the water temperature from 70°F to 32°F. This requires (70° - 32°) x 1 BTU/Lb/°F = 38 BTU's.

2.   Secondly, 32° water must be transformed into 32° ice. This phase change requires 144 BTU's.

3.   Thirdly, the 32° ice must be reduced in temperature to 10°F. This requires (32° - 10°) x 0.5 = 11 BTU's.

The total energy required equals 38 + 144 + 11 = 193 BTU's. (See Figure 2)

This example shows the significance of the water content of a food item.

The higher the water content, the more energy it takes to freeze. Further discussion of the freezing process is not of benefit here. The reader is referred to "Frozen Food Technology", C.P. Mallett.

2.4.1   *Water that Doesn't Freeze (!)*

Later in Chapter 4 – 4.2.3 we can see that in calculating the energy required to freeze a food item we will take into account the percentage of water that is ***not*** frozen. It is beneficial to the reader to know why.

The initial freezing temperature of an aqueous solution decreases as the concentration of solute increases. For example, the freezing temperature of a 100% sugar solution is 0.5°F below that of pure water. It follows that the freezing rate is influenced by the percentage of sugar solution in the food. Since most foods also contain sugars or proteins dissolved in water, freezing is also controlled by the percentage of solute at a specific temperature. For instance, beef with 74.5% of water at 0°F has 12% of its water not frozen; at -40°F 10% has not been frozen.

Kerr and Reid in "Thermodynamics and Frozen Foods" cover this topic sufficiently. (See Tables 4, 5, 6)

Table 4 ENTHALPHY OF FROZEN MEATS[1]

Product	Water Content (% by wt)	Mean Spec. Heat 40° to 60°F		-40	-20	-10	-5	0	5	10	15	18	20	22	24	26	28	30	32
											Temperature, °F.								
Beef, lean, fresh	74.5	0.84	Enthalpy BTU/lb.	0	9	15	18	21	24	27	32	35	38	42	48	57	74	119	131
			% Water unfrozen	10	10	11	12	12	13	15	18	20	22	24	28	37	48	92	100
Beef, lean, dried	26.1	0.59	Enthalph BTU/lb.	0	9	14	17	20	24	28	31	-	33	-	36	-	38	-	40
			% Water unfrozen	96	96	96	97	98	99	100	-	-	-	-	-	-	-	-	-
Cod	80.3	0.88	Enthalpy BTU/lb.	0	10	15	18	21	24	28	33	36	39	43	48	56	73	123	139
			% Water unfrozen	10	10	10	11	12	13	14	16	18	20	22	26	32	45	88	100
Haddock	83.6	0.89	Enthalpy B'TU/lb.	0	9	15	18	21	24	28	33	36	39	43	48	56	73	127	145
			% Water unfrozen	8	8	9	9	10	11	12	14	15	17	19	23	29	42	86	100
Perch	79.1	0.86	Enthalpy BTU/lb.	0	9	14	17	20	23	27	32	35	38	42	46	53	68	117	137
			% Water unfrozen	10	10	11	11	12	13	14	16	17	19	21	24	30	41	83	100
Chicken, young rooster	76	0.85	Enthalpy BTU/lb.	0	9	14	17	20	23	27	32	36	39	43	49	57	72	122	132
Chicken, dried young rooster	23.5	0.56	Enthalpy BTU/lb.	0	9	14	16	19	21	24	27	28	29	30	31	32	-	35	36
Veal, fresh	76.5	0.85	Enthalpy BTU/lb.	0	10	15	18	21	24	28	33	36	39	43	48	57	72	123	133
Venison	73	0.84	Enthalpy BTU.lb.	0	9	14	17	20	24	28	33	37	41	46	52	61	78	119	127

[1] Above -40°F.

Adapted from Riedel (1956, 1957A).

Solid-Liquid Equilibria

TABLE 5

PERCENT FROZEN WATER IN ICE CREAM AT VARIOUS TEMPERATURES[a]

% Water frozen to ice	°C (°F)	% Water frozen to ice	°C (°F)
0	-2.5 (27.55)	40	-4.2 (24.40)
5	-2.6 (27.35)	45	-4.6 (23.56)
10	-2.7 (27.05)	50	-5.2 (22.62)
15	-2.9 (26.78)	55	-5.9 (21.42)
20	-3.1 (26.40)	60	-6.8 (19.79)
25	-3.3 (26.04)	70	-9.5 (14.99)
30	-3.5 (25.70)	80	-14.9 (5.14)
35	-3.9 (25.03)	90	-30.2 (-22.29)

Composition: fat, 12.5%; serum solids, 10.5%; sugar, 15%; stabilizer, 0.30%; water, 61.7%

[a]From Liska and Rippen (24). Presumably based on initial water content.

TABLE 6

PERCENT FROZEN WATER[a] IN VARIOUS FOOD MATERIALS

°C (°F)	Lean meat[b] (74.5% H_2O)	Haddock[c] (83.6% H_2O)	Egg White[d] (86.5% H_2O)	Egg Yolk[d] (50% H_2O)
0 (32)	0	0	0	0
-1 (30.2)	2	9.7	48	42
-2 (28.4)	48	55.6	75	67
-3 (26.6)	64	69.5	82	73
-4 (24.8)	71	75.8	86	77
-5 (23.0)	74	79.6	87	79
-10 (14.0)	83	86.7	92	84
-20 (-4)	88	92.6	93	87
-30 (-22)	89	92.0	94	89
-40 (-40)	—	92.2	—	—

[a](grams ice/grams of total initial water) x 100

[b]Riedel (41)

[c]Riedel (40)

[d]Riedel (42)

3. MECHANICAL VS. CRYOGENIC FREEZING

3.1 Using Cryogenic Food Freezers Effectively

There are many refrigeration processes available to the food processing industry. However, only two are normally of importance. The first one is a closed loop mechanical refrigeration system using Ammonia or Freon with a compression, expansion and cooling loop to achieve a low temperature environment. The second one is a so called expendable refrigeration process and uses liquid nitrogen or carbon dioxide. These processes are known as cryogenic freezing systems. Here, the refrigerant is injected into the freezer as a liquid, allowed to change phase to a gas (liquid, solid, gas for carbon dioxide) during which enormous amounts of energy are released resulting in very low temperatures. With this process, the refrigerant is used only once and vented into the atmosphere by mechanical means.

> *Cryogenic food freezing differs significantly from mechanical ammonia or Freon blast freezing. The method of measuring the final product end temperature is different and it is therefore described in the following paragraphs.*

3.1.1 Mechanical Freezing Systems

Due to a much higher freezer temperature, mechanical freezers freeze very slowly (See Fig. 4) and evenly through the product. The temperature difference between the final required product temperature and the freezer is minimal. Consequently, when a product exits a mechanical freezer, it is at its final temperature. If this temperature is not acceptable, the dwell time in the freezer is the only variable that can be altered to achieve the desired product temperature.

Finally, mechanical freezers (Spirals or linear belt) rely heavily upon cold air circulation through each belt level, and product placement can not exceed 80% for an effective operation.

3.1.2 Cryogenic Freezing Systems

Cryogenic freezing processes are the exact opposite of mechanical freezing processes and require a different technique of belt loading and product temperature measurement.

> *A product frozen cryogenically freezes rapidly from the exterior of the food product to a temperature well below the desired product end temperature, while the core of the product trails behind. The final product temperature is measured several minutes*

after the EQUILIBRATION period has taken place!!

During the equilibration period, the cold that has been impinged into the outer layers of the product will be absorbed by the core by means of thermal conductivity. This process takes several minutes to complete and of course, must be carried out in a cold environment. Otherwise the cold in the outer layers will be warmed by the environment rather than equilibrate with the warmer core.

Thermal Conductivity is a property of a material that expresses the heat flux which is a measure of the rate (Also see Paragraph 3.1.4) at which heat is transferred through a given thickness of a product if a certain temperature gradient exists over the material. It is very difficult to measure thermal conductivity and it requires a steady state situation and can only be performed under controlled laboratory conditions.

For our general understanding of the freezing process however it is helpful to have knowledge of this phenomenon. Products with higher thermal conductivity value will generally freeze more rapidly. For instance, raw unfrozen shrimp has a thermal conductivity of 0.33 BTU/(hr)(ft^2)(°F/ft). And unfrozen beef has a thermal conductivity of 0.26 BTU/(hr)(ft^2)(°F/ft). We can see that heat can be removed through a given thickness of shrimp about 27% faster than beef of the same thickness.

3.1.3   Product Temperature Equilibration

◊ The temperature of the product as it leaves the cryogenic tunnel is not its final temperature. This final temperature will not be known until 5 to 10 minutes have lapsed and full equilibration has taken place.

It follows; of course, that the core temperature of the product as it leaves the tunnel will be higher in temperature than the required temperature and further freezing of the product is to take place by thermal conductivity. If the core temperature of the product is measured directly after it leaves the tunnel, an incorrect assumption can be made about the set-point temperature of the tunnel. **Lowering the set point temperature will increase the specific LN2 consumption dramatically.**

Cryogenic freezing operates with a huge temperature difference between the final product temperature and the freezing chamber, often as much as 120°F. This difference assisted by high turbulence results in a very fast temperature reduction of the food item. Consequently, dwell times are very short but a period of temperature equilibration is required to affect the desired product core temperature. If the product temperature is not acceptable at a specific set-point temperature, two adjustments are available to lower the product temperature:

1.   Belt speed (Dwell time) – Reduce the belt speed.

2.   Rate of turbulence – Increase the fan speed if possible.

If the freezer's temperature has to be lowered below the designed set point temperature, often -160°F, to get the product frozen to an acceptable temperature, the answer is simple. The tunnel is too small! (Also see 3.1.8)

3.1.4   Freezing Rate

The definition of freezing refers to the reduction of the temperature of a foodstuff from its original temperature to a temperature below its freezing point. As so well described in "Quick Frozen Foods", Global frozen food almanac (1991) and many other publications, freezing of food stuffs must take place rapidly to keep the ice crystals small as not to puncture the cell structure (See Figure 5). That's why fast freezing produces a superior product when thawed and hence, is simply a function of the freezing rate expressed as (W).

$$\text{Also, } W = r_1/t$$

Where (r_1) is the distance from the items surface and the items core and (t) is the freezing time to transfer T1 to T2.

Cryogenic freezers are designed with a (W) rating of at least 2 inches/hr.

The freezing rate in a specific freezer is, as described further in this book, dependent on how the LN2 is injected into the freezer, and the air turbulence in the freezer. Products should never be stacked during the initial crust freeze.

3.1.5   Product Belt Loading

How the product is arranged on the conveyor belt is critical to the efficient operation of a cryogenic freezer. However, we have to differentiate between a "SPRAY COUNTER FLOW" tunnel and "CONVECTION" tunnel. Unlike a mechanical freezer, where about 40% of the belt surface should remain open, the turbulent cold gas within the Spray Counter flow tunnel does not have to pass through the belt to affect the freeze. Therefore, the belt is loaded with only a minimal gap between the products and as close as possible to the side of the conveyor belt. The objective is to have the maximum product surface covered by the LN2 spray when it passes through the spray zone of the tunnel.

◊ A Convection tunnel on the other hand requires some level of turbulence through the belt and around the product. A belt loading of 80% (product shape and size will eventually determine the correct spacing) is generally recommended.

3.1.6   The Belt Speed

The belt speed may be adjusted to vary the product's dwell time or residence time in the freezer to influence the product's end temperature. The longer the product stays in the freezer the lower its final temperature. Too long of a dwell time at a specific set point temperature may cause product damage and also a waste of LN2 and should be avoided. Generally, the belt speed is determined by the required *production rate* in pounds per hour and is the first parameter used during the initial design stage.

3.1.7   Turbulence Fans

Adjusting the turbulence fans will vary the rate of turbulence or cold air velocity within the tunnel, if the freezer is equipped with a variable speed controller. The air circulation in a cryogenic tunnel is crucial to its effective operation. As a general rule, the turbulence must be as high as possible, around 1500 ft/min. It is sometimes necessary to reduce the turbulence to prevent small and delicate products from becoming air born and thrown off the belt. A remedy to this is the installation of guide rails along both sides of the conveyor. Now we can operate the circulating fans at a relative higher speed without the product being blown off the belt.

Properly shaped product guide rail.

Wrongly shaped product guide rail

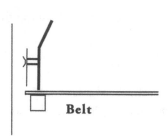

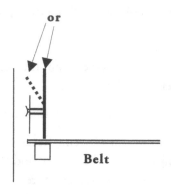

3.1.8   Set Point Temperature

The set-point temperature is the actual temperature in the freezer and is adjustable from 0 to -160°F. Although in theory, with a LN2 freezer, the temperature can be adjusted as low as -250°F, but this is not recommended to avoid damaging the freezer's insulated body panels.

The operating temperature of the tunnel displayed on the temperature controller is actually a relative temperature and is not the absolute temperature in the tunnel. The temperature in the cryogenic zone of a spray counter flow tunnel can be as low as -320°F. The recommended set point temperature of -160°F is the *average temperature* of the entire tunnel. The placement of the temperature sensor plays a roll in the overall average tunnel temperature. The closer the temperature probe is installed to the cryogenic spray zone, the "warmer" the tunnel's average operating temperature is.

In a convection tunnel, where there is no cryogenic spray zone, the average temperature in the tunnel is very much the temperature in the entire tunnel.

In a spiral freezer the set point temperature is the temperature in the entire freezer.

Cryogenic tunnels can be and are, configured in many different ways and the temperature at which they operate can differ accordingly. Set point temperatures between (-90°F) and (-160°F) will suffice for most applications; **assuming the freezer is sized correctly.** Continuous operation at temperatures lower than this range can cause an inefficient operation and result in wasted energy, LN2. When adjusting the temperature, lower it by 5° intervals and the results monitored every 5 minutes or so.

3.1.9 *The Exhaust Gas Temperature*
This measurement is a good indication if the freezer is adjusted and/or sized correctly.

The Spray Counter Flow Tunnel
Here, the exhaust gas temperature exits at the entrance of the tunnel and should be as close as possible to the desired product equilibrated temperature. In general this temperature is about 20 degrees below the desired product equilibrated temperature for a raw product entering the tunnel at approximately 45°F. It is not uncommon, when a hot cooked product is being frozen, to see an exhaust gas temperature equal to the desired product temperature.

The Convection Tunnel
Is always fitted with exhaust plenums on both sides of the tunnel. Here, there is little correlation between the product end temperature and the exhaust vapors. The temperatures on both ends of the tunnel should not be too close to the freezer's set point temperature. At a set point temperature of -160°F, generally the exhaust should be around -80°F. Of course, this is only possible if the tunnel has the correct length for a specific production rate. As with the spray tunnel, processing a hot product will influence this temperature. The design of a convection tunnel proves to be a greater challenge for the designer than with the spray tunnel. Since the temperature of the exhaust vapors determine largely how efficient the freezing process will be, not having full control of this temperature will have a negative affect on the efficiency of the freezer.

In the following chapters, more than once, shall I emphasize that the exhaust-gas temperature-management is the most important function for the operator to watch during the freezing process.

Figure 7 depicts clearly how the exhaust gas temperature influenced the efficiency of the freezing system. At (+20°F), the BTU's in one pound of LN2 is 80.5 + 83.1 = 164 BTU's. However, if the same freezer has an exhaust temperature of (-20°F), the available BTU's are 80.5 + 73 = 153.5 BTU's/Lb. or 7% less.

Figure 4 – Freezing curves of fast and slow freezing

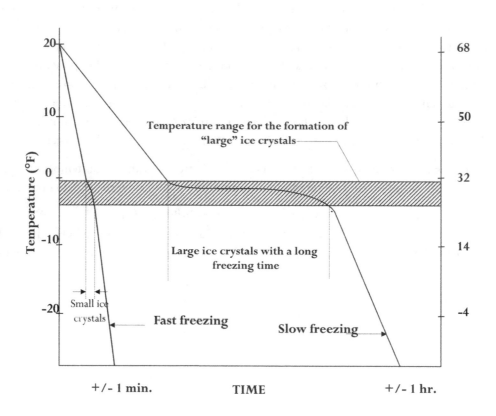

Fig. 5 – SHRIMP TISSUE SLIDES

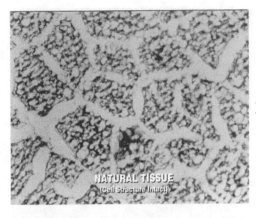

Fresh Tissue

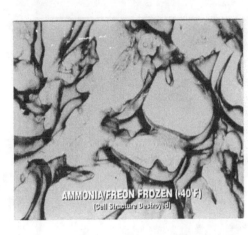

The Ammonia frozen shrimp revealed considerable cell tissue damage when thawed.

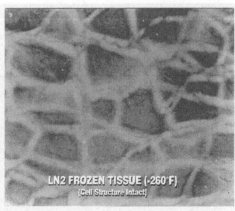

The LN2 frozen shrimp revealed no cell tissue damage when thawed.

4.   CALCULATING THE LN2 CONSUMPTION

4.1   COOLING POWER FROM LIQUID NITROGEN

4.2   Operational factors to consider

4.3   CALCULATING LIQUID NITROGEN CONSUMPTION

Figure 7   Refrigeration Capacity of LN2 @ 20psig as a function of the Exhaust Gas Temp.

4.1 COOLING POWER FROM LIQUID NITROGEN

This handbook will focus primarily on the use of LN2 as the refrigerant. (See also Chapter 14 for LCO$_2$) In cryogenic freezing with LN2, the cooling power comes from:

1. **The Heat of Vaporization.** This is,
   The cooling power generated when liquid nitrogen undergoes a phase change to gaseous nitrogen.
2. **The Sensible Heat.** This is,
   The cooling power generated as the gaseous nitrogen is heated by its surroundings.

4.1.1 Heat of Vaporization

Nitrogen in its liquid state has less energy than nitrogen in its gaseous state. What this means is that for nitrogen to change from a liquid state to a gaseous state, energy is required. This energy comes from an outside source in the form of heat. The amount of heat, in BTU's, necessary to change one pound of liquid nitrogen to gaseous nitrogen is known as the "HEAT OF VAPORIZATION". A phase change takes precedence over a rise in temperature. This means that a liquid will totally change into a gas *without* changing its temperature.

4.1.2 Sensible Heat

Nitrogen in its gaseous state can absorb heat and rise in temperature. The amount of heat required to change the temperature of one pound of nitrogen one degree Fahrenheit is called the "SPECIFIC HEAT". The cooling power available in the gaseous form is known as the "SENSIBLE HEAT". The equation relating the sensible heat to the specific heat is expressed as:

$$q = m \times Cp \times \Delta T$$

Where:
q = Sensible Heat (BTU)
m = mass of the gas (LB)
Cp = Specific Heat (BTU/LB/°F)
ΔT = Temperature differential (°F)

Now that we are familiar how liquid nitrogen is able to freeze, let us examine some additional factors affecting this cooling power.

4.1.3   Pressure

Table 2 lists the heat of vaporization and the heat capacity for nitrogen at varying pressures. Examination of this table shows that as the pressure increases the heat of vaporization decreases. **This is why, in cryogenic freezing applications, the storage tank pressure is kept to a minimum.** With a lower tank pressure we will maximize the amount of cooling power available from the phase change of liquid nitrogen. Further examination of the table shows that the specific heat (Cp) does vary with pressure, but not as significantly as the heat of vaporization does.

4.1.4   Boiling Point Temperature

Table 2 shows that the boiling point temperature (T_{BP}) of liquid nitrogen is directly proportional to its pressure. Therefore, as the pressure increases, the boiling point temperature will increase. The sensible heat (q) shows that it is indirectly proportional to the tank pressure. The following equation is used to determine the total energy derived from one pound of liquid nitrogen in a freezing tunnel.

$$\mathbf{Q_{Total} = m(\Delta T \times Cp) + H_{vap}}$$

4.1.5   Exhaust Gas Temperature

It is apparent that the only variable here is the temperature differential. The temperature differential is simply:

$$\mathbf{\Delta T = T\ initial - T\ exhaust}$$

Where:
ΔT = Temperature differential, which is
T initial = Nitrogen boiling point at system tank pressure
T exhaust = Exhaust gas temperature

The initial temperature of the gas is in general the boiling point of the nitrogen at a specific pressure (remember that we normally start with a liquid nitrogen which vaporizes into a gas at that boiling point); the only variable here is the exhaust gas temperature. The equations have shown us that the higher the exhaust gas temperature, the greater the amount of energy we can extract from the nitrogen.

4.1.6   Temperature Differential

The temperature differential is simply the nitrogen's boiling point temperature T_{BP}, minus the temperature of the exhaust gas as it exits the freezer.

4.1.7   Time

An important factor in maximizing the cooling effect of liquid nitrogen is time. The longer the nitrogen (vapor) is in contact with the product to be frozen or cooled the greater is the cooling effect and the greater amount of energy is transferred to the product. It is for this reason that the freezing tunnels are best designed as long as possible. Short tunnels are very inefficient and operate with low exhaust gas temperatures. In Figure 7 we see that a low exhaust temperature reduces the refrigeration capacity of liquid nitrogen. A short tunnel, 10 to 15 feet, running with a set point temperature of -160°F will have a high internal pressure which almost certainly will result in a very low exhaust temperature, -100°F or colder.

Do not operate tunnels shorter than 20 feet with set point temperatures lower than -60°F if freezing economy is a factor.

4.1.8   *Losses Due to a High Tank Pressure*

Figure 6 illustrates how a high tank pressure will contribute to an inefficient freezer. For instance, at a tank pressure of 20 psig, the LN2 released in the freezer will flash off 7.7% of nitrogen vapors entering the spray zone. In a spray tunnel, these vapors will not come in direct contact with the product. It only will increase the vapor pressure inside the freezer making it difficult to balance the exhaust.

4.2   Operational factors to consider

Before calculating the liquid nitrogen consumption one should be aware of operational factors that will influence the designed of the tunnel and consequently will affect the overall efficiency of the tunnel. Such as:

4.2.1   *Product Refrigeration Requirement and Type of Product*

This is the refrigeration requirement as determined in (Paragraph 4.3) which should indicate if we are freezing or just cooling the product and the type of product to be processed. If we are just cooling the product, the product end temperature is at best 32°F, we could design the tunnel as an convection type tunnel with a less complicated circulation arrangement. In our example, we have chosen ice cream to be frozen to a temperature well below 32°F. With this in mind, we know that we are *freezing* ice cream and the type of tunnel should be a convection tunnel. Ice cream can be processed as small novelty type product or they can be processed in 1 gallon containers. In either case, we should select a belt with a large mesh such as a B18-16-16 or larger to promote good circulation around the product. If the small novelty ice cream is light in weight, 1 to 10 grams, we should use a circulation system designed with a variable speed controller to prevent too much circulation within the tunnel.

Although ice cream can be frozen in a convection *or* spray tunnel, a confection tunnel is the better choice. Exposing un packaged ice cream to a LN2 spray can result in some surface cracks. Hence, most ice cream products, packaged or unpackaged, are frozen in a confection type freezer. Unpackaged ice creams are generally very solid at 20°F and can be further processed without losing its shape.

4.2.2   *The Product is Packaged*

Product packaged before the freezing process will increase the liquid nitrogen consumption and if possible should be avoided. The liquid nitrogen consumption can increase by a factor of 2 to 4 depending on the thickness of the packing material. If a packaged product is to be frozen, the maximum air circulation within the tunnel and around the individual product is required. Make sure that at least ½" of spacing is provided around the product.

4.2.3   Not All the Moisture in the Product is Frozen

The physical constituents of the product may be such that not all of the moisture in the product will be frozen. (See tables 3, 4 and 5) If from available data the percent unfrozen water is known, the calculations should be adjusted accordingly as illustrated in Example (A & B).This is essential if accurate liquid nitrogen consumption data is required. As illustrated it may over state the consumption by as much as 10%.

Example A

Calculate the heat to be removed from one pound of Fillet of Cod to be frozen from 40°F to -10°F.

Starting temperature (T1)	40°F
Specific Heat Above freezing	0.84 BTU/LB/°F
Freezing Point	28°F
Water content	80%
Specific Heat Below freezing	0.44 BTU/LB/°F
Desired end temperature (T2)	-10°F

Conventional method of calculating LN2 usage,

H1 = 1x(40° - 28°)x 0.84	= 10.08 BTU
H2 = 1x 144 x 80%	= 115.20
H3 = 1x {(28° - (-10°)}0.44	= 16.72
Total Heat removed	= 142.00 BTU/LB

Calculation with 10% of the water not frozen (Table 3),

H1 = 1x(40° - 28°)x 0.84	= 10.08 BTU
H2 = 1x (144 x 80%) - 10%(un-frozen)	= 103.68
H3 = 1x {(28° - (10°)}0.44	= 16.72
Total Heat removed	= 130.48 BTU/LB

The difference is 11.52 BTU or 9 %. This is considerable when thousands of pounds of product are frozen per day.

4.2.4   Cool Down Requirements

This is the energy or the quantity of liquid nitrogen used to cool-down the tunnel to the desired operating or set-point temperature. The cool-down process should begin as closely timed to the actual production start as possible. In general it should not take more than 15 minutes to get the freezer to the set-point temperature. *Adjust the exhaust blower to the minimum power setting. Do not overlook this!*

4.2.5   Fixed Operating Load

This is the minimum amount of nitrogen required to maintain the freezer at the set-point temperature to overcome the heat input from the circulating fans, and the heat leak through the walls.

Although all of the above factors can be mathematically calculated, this method is seldom used. Too many factors are difficult to verify and are greatly influenced by the dedication of the operator.

A very accurate method is to compare the amount of product processed through the freezer within a specific time, say 30 working days, and compare it with the amount of liquid nitrogen used within that same time period.

This method is accurate however; it will not reveal were the losses have occurred if the results indicate that excessive amounts of LN2 have been used. Thorough examination of the freezer and the method of operation can usually reveal where the excessive use of LN2 originated. See Chapter 11, "Why the freezer is using too much LN2".

Now proceed to paragraph 4.3 to complete the calculation.

4.3 CALCULATING LIQUID NITROGEN CONSUMPTION

The Specific Heat of a food item is the,

"Amount of heat that must be removed or added to change the temperature of one pound of the food item one degree fahrenheit".

It is expressed in BTU/Lb/°F. Each food has a specific heat value *above* its freezing point and a different (lower) value *below* its freezing point.

The freezing points of most foods are within a few degrees of 32°F, which is the freezing point of water. When foods are frozen, it's actually the water content of the food that is frozen. As we have seen from previous chapters, it requires the removal of 144 BTU to freeze one pound of water (change 32°F water to ice of 32°F, just a change of state *without* a change of temperature).

To calculate the LN2 consumption to freeze a food item, three major heat loads must be combined. These are:

1. Heat load of the product
2. Equipment cool down requirement
3. Fixed operating load

To calculate the heat to be removed from the food item requires three steps:

1. The heat that must be removed from the product to reduce its temperature just to its freezing point.
   H_1 = weight x (Sp.Heat. above Fr.) x (T1 – Product Freezing Temp.)
2. The heat that must be removed to freeze the product (freezing the H_2O in the product) also called, Latent Heat.
   H_2 = weight x 144 x % H_2O/100
3. The heat that must be removed from the product to reduce the temperature further to the desired T2 or final temperature.
   H_3 = weight x (Sp.Heat below Fr.) x (Fr.Temp. of the product – T2)

Example B

Calculate the heat that must be removed to freeze ice cream from 45°F to -22°F.

Ice cream physical characteristics from chart:

Weight	= 1 pound	
Average freezing point	= 22 to 29°F.	(From tables)
% H_2O	= 58 to 66	(From tables)
Specific Heat above Fr.	= 0.80 BTU/LB/oF.	(From tables)
Specific Heat below Fr.	= 0.45 ,,	(From tables)
Latent Heat	= 96 BTU/lb.	(From tables)

The total heat that must be removed will be:

From 45° to -22°F, **Using Table 4**

H1 = 1 x 0.80 x (45° - 22°) = 18.40 BTU = 18.40 BTU

H2 = 1 x 144 x 66/100 = 95.04 1 x 90%(144 x 66/100) = 85.54

H3 = 1 x 0.45 x (22 – (**-22°**) = 19.80 1 x 0.45 x (22° –(-22°) = 19.80

Total Heat Removed **133.24 BTU/Lb.** **Total Heat Removed 123.74 BTU/Lb.** (8% less)

Example C

Now calculate the heat that must be removed from a pound of cod fillet to freeze from 40°F to a reduced temperature of -10°F.

Weight	1 pound
Average Freezing Point	28°F
%H_2O	80%
Starting temperature T1	40°F
Specific Heat Above Freezing	0.84 BTU/LB/°F
Specific Heat Below Freezing	0.44 BTU/LB/°F
Required final temperature T2	-10°F
Latent Heat	Not known

Using Table 4

H1 = 1 x 0.84 x (40 – 28) = 10.08 BTU/Lb. = 10.08 BTU/Lb.

H2 = 1 x 144 x 80/100 = 115.20 1 x 90%(144 x 80/100) =103.68

H3 = 1 x 0.44 x (28 – (-10) = 16.72 1 x 0.44 x (28° - (-10°) = 16.72

Total Heat Removed **142.00 BTU/Lb.** **130.48 BTU/Lb.** (9% less)

4.3.1   Calculating how much LN2 is Required

First determine the pressure in the storage tank. At a tank pressure of 20 PSIG, LN2 has a Heat of Vaporization of 80.5 BTU/lb. Use (Table 2) In addition, nitrogen vapors at that pressure have a heat capacity (Cp) of 0.255 BTU/Lb/°F.

The GN2 has an energy value limited by the temperature of the vapors as they *exit* the freezer. If the vapors are exhausted at $-30°F$ for instance, the sensible heat energy is calculated as follows:
S.H. = ΔTgas x Cp = (-305.7°– (-30°) x 0.255 = 70.3 BTU

The total energy from one pound on LN2 is now, Q_{Total} = 70.3 + 80.50 = 150.8 BTU/lb.

The LN2 consumed in Example B can now be calculated as, 133.24/150.8 = 0.88 or 123.74/150.8 = 0.82 Lb/Lb. **(About 8% less)**

4.3.2   Cool Down Requirements

This is the energy or the quantity of liquid nitrogen used to cool-down the freezer to the desired operating temperature. The cool-down process should begin as closely timed to the actual production run as possible. In general it should not take more than 15 minutes to get the freezer to the desired running or set-point temperature. *Adjust the exhaust blower to the minimum power setting during this step. Do not overlook this!*

It is possible to get a relative accurate value for the cool down requirement. Before starting the freezer, record the level of LN2 in the storage tank in inches of water column. Now, turn on the freezer and take a second reading when the tunnel is at the set point temperature, again in inches of water column. The difference between the two readings is the LN2 consumed to cool down the freezer. A conversion chart is provided by the LN2 supplier with the tank, to convert inches of water column to actual LN2 in gallons or pounds.

> *CAUTION: Make sure to get a conversion chart that corresponds to the exact tank pressure used while taking the readings.*

4.3.3   Fixed Operating Load

This is the amount of liquid nitrogen required to maintain the freezer at the set-point temperature to overcome the heat input from the circulating fans (A) the running conveyor belt (B) and possible heat leak through the walls (C).

A.   Determine the HP of the freezer's circulating fans. For example, 8 fans of 1 HP each. These accumulatively produce a heat load of 8 x 2544.5 BTU/hr. = 20,356 BTU/hr.

B.   A 26 inch wide conveyor belt introduces about 120 BTU/ft/hr. into the tunnel. It follows that this tunnel with a linear belt length of 20 feet introduces 20 x 120 = 2400 BTU/hr. into the tunnel. A 20 feet tunnel with a 36 inch wide belt introduces (36/26 x 120) 20 = 4,320 BTU/hr. into the tunnel.
A 40 feet tunnel with a 48 inch wide belt introduces (48/26 x 120) 40 = 8,862 BTU/hr into the tunnel, etc.

C.   The walls of the freezer have, for instance a surface area of 250 square feet and are insulated with 4 inches of high density Polyurethane with a heat leak number of 59BTU/ft./hr.This gives a total load of 250 x 59 = 14,750 BTU/hr.

NOTE: Heat leak from radiation through the walls of a *new* tunnel can be neglected. It follows that the steady state losses for this freezer are the sum of A, B and C or 37,506 BTU/hr.

Theoretically, this requires 37,506 / 150.8 or 248.7 Lbs. of LN2 per hour.

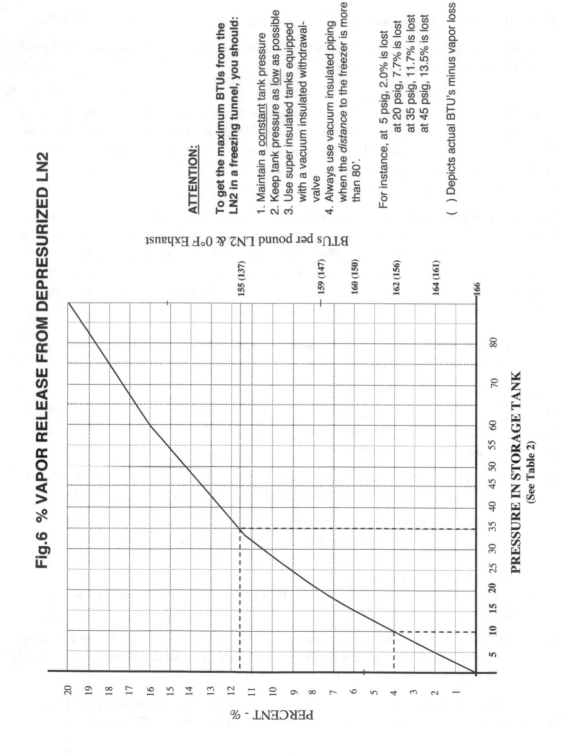

Fig.6 % VAPOR RELEASE FROM DEPRESURIZED LN2

BTUs per pound LN2 & 0°F Exhaust

ATTENTION:

To get the maximum BTUs from the LN2 in a freezing tunnel, you should:

1. Maintain a _constant_ tank pressure
2. Keep tank pressure as _low_ as possible
3. Use super insulated tanks equipped with a vacuum insulated withdrawal-valve
4. Always use vacuum insulated piping when the _distance_ to the freezer is more than 80'.

For instance, at 5 psig, 2.0% is lost
at 20 psig, 7.7% is lost
at 35 psig, 11.7% is lost
at 45 psig, 13.5% is lost

( ) Depicts actual BTU's minus vapor loss

155 (137)
159 (147)
160 (150)
162 (156)
164 (161)
166

PERCENT - %

PRESSURE IN STORAGE TANK
(See Table 2)

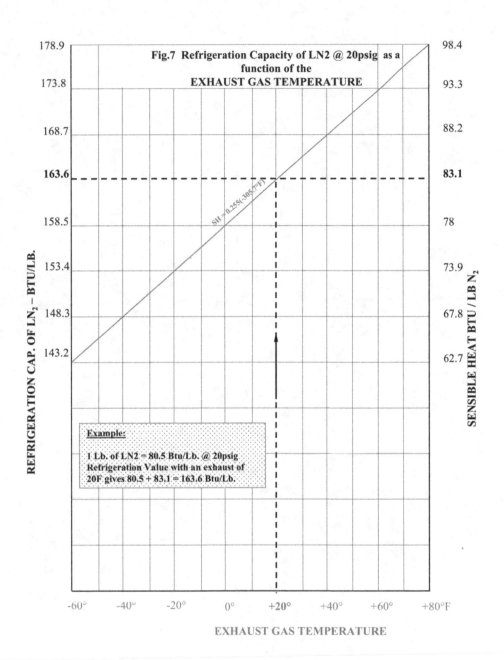

Fig.7 Refrigeration Capacity of LN2 @ 20psig as a function of the EXHAUST GAS TEMPERATURE

Example:

1 Lb. of LN2 = 80.5 Btu/Lb. @ 20psig Refrigeration Value with an exhaust of 20F gives 80.5 + 83.1 = 163.6 Btu/Lb.

4.4.0 *Formulas for calculating the Specific Heat and Latent Heat of Fusion when only the water content of the product is known*

1. To calculate the Specific Heat *Above* Freezing:

   Sp.Ht. Above freezing = $(\%H_2O)/100 + 0.42 (1 - \%H_2O/100)$

2. To calculate the Specific Heat *Below* freezing:

   Sp.Ht. Below Freezing = $0.5 (\%H_2O/100) + 0.42 (1 - \%H_2O/100)$

3. To calculate the Latent Heat of Fusion:

   $L = 144 (\%H_2O/100)$

Example: Find the specific heat above and below the freezing point and the latent heat of fusion for a product having a moisture content of 80%.

1. Sp Ht. AF.= $80/100 + 0.42 (1 - 80/100) = 0.88$ BTU/Lb/°F.

2. Sp. Ht. BF.= $0.5 (80/100) + 0.42 (1 - 80/100) = 0.48$ BTU/Lb/°F

3. $L = 144 (80/100) = 114$ BTU/Lb.

4.4.1 *Calculating the Freezing Cost*

For COOLING to a temperature above the product's freezing point, use:

$X = D (A - B)$

For FREEZING to a temperature below the product's freezing point, use:

$X = D (A - C) + E (C - B) + F$

FREEZING COST

$Z = Y (X/G)$

Where A = Starting temperature (T1)

B = Final equilibrated end temperature (T2)

C = Freezing point of the product

D = Product Specific Heat Above Freezing in Btu/Lb/°F

E = Product Specific Heat Below Freezing in Btu/Lb/°F

F = Product Latent Heat of Fusion in Btu/Lb

G = Total Heat of Vaporization of LN2 (See Table 2, Q_{Total})

X = Btu's required to freeze or cool per pound of product

Y = Cost of the LN2 *per pound*

Z = Freezing cost per pound of product processed

4.4.2   All the Formulae You Need to Determine the LN2 Consumption to Freeze a Product

4.4.2.1 Chaim's Formula

Q_{Prod} = QA (T1-TQ) + L +QB (TF – T2) = BTU/Lb of product.

Where:

Q_{Prod} = Individual product heat transfer rate in BTU's/Lb.
QA   = Specific Heat of the product ABOVE the freezing point in (BTU's/Lb.)
QB   = Specific Heat of the product BELOW the freezing point in (BTU's/Lb.)
L   = Latent Heat of fusion of the product in BTU's/Lb.
T1   = Initial product temperature in °F
T2   = Final *equilibrated* product temperature in °F

If we don't know QA, QB and L then proceed as follows:

QA   = 0.5(% Fat) + 0.35(% Solid) +1(% H_2O) = …BTU's/Lb/°F
L   = 144 (% H_2O) = … BTU's/Lb.
QB   = 0.5(% Fat) + 0.35(% Solid) + 0.5(%H_2O) =…BTU's/Lb/°F

4.4.2.2 Calculating the Specific Heat and Latent Heat of Fusion when only the water content of a product is known.

1.   Specific Heat ABOVE freezing: (%H_2O/100) + 0.42(1 - %H_2O/100)

2.   Latent Heat of Fusion: 144 (%H_2O/100)

3.   Specific Heat BELOW freezing: 0.5 (%H_2O/100) + 0.42 (1 - %H_2O/100)

4.4.3   Calculating the Energy Available from One Pound of LN2

$$Q_{Total} = H_{vap}. + \text{Sensible Heat in BTU/Lb.}$$

1.   Determine the Heat of Vaporization (H_{vap}) of the LN2 at the operating tank pressure. (See Table 2)
   At 10 PSIG the H_{vap} = 82.60 BTU/Lb., at 15 PSIG the H_{vap} = 81.50 BTU's/Lb. at 20 PSIG H_{vap} = 80.50 BTU's/Lb.

2.   Determine the LN2 boiling point (T_{BP}) at that pressure.
   For instance, at 20 PSIG T_{BP}) = -305.7°F

3.   Determine the exhaust vapor (E_{vap}) temperature at the freezer.

4.   Determine the Heat Capacity (Cp) of the nitrogen vapors for the specific LN2 pressure from Table 2.

5.   We can now determine the Sensible Heat (SH) of the LN2 with:
   SH = ΔTvapors x Cp
   Where, ΔTvap. = (T_{BP} – Evap.)
   Evap. = Exhaust vapor temperature measured at the beginning of the freezer's exhaust plenum, about 12" above the belt.

4.4.3.1 Determine the LN2 Used To Freeze a Product
(Also see paragraph 4.3.1)

$$Q_{Prod} / Q_{Total} = \text{Lb of LN2 / Lb of prod.}$$

Where:

Q_{Prod} = Individual product heat transfer rate in BTU/Lb.

Q_{Total} = The energy available from the LN2 heat of vaporization plus the sensible heat of the vapors.

5.   HOW TO CALCULATE THE FREEZER SIZE

5.0 An overview of design fundamentals

Liquid Nitrogen (LN$_2$) is a highly dynamic fluid, prone to phase changes, which are difficult to detect, so special care must be taken during process engineering design of the freezers *and* the systems that contain the LN2 during the freezing process.

A properly designed cryogenic freezer will operate for 20-plus years and can be virtually forgotten about. An improperly *sized* cryogenic freezer will never be forgotten because the production manager will constantly attempt to correct the performance deficiencies, and the plant manager will constantly question the high liquid nitrogen utilization rate, which affects the freezing cost. In addition, a poor vacuum of the LN2 tank, faulty filling procedures, too small piping, or all of the above often causes excessive losses.

It is astonishing to see that not many cryogenic freezers installed over the years have been sized properly. Many installations producing thousands of pounds of product are actually inefficient and use excessive amounts of liquid nitrogen. Often, the original freezer was sized for a much smaller capacity and, at times, a different product as well. If the operators had some insight of tunnel design, enormous amounts of money could have been saved. This chapter will guide the reader through a process to avoid these mistakes.

Here are the steps we can easily follow and henceforth understand which parameters are necessary to design a freezer to freeze a specific product.

1. **Determine the exact nature of the product. (We assume the product is unwrapped)**

    What kind of product is it?

    Is the product raw or cooked?

    Is the product to be frozen or chilled?

2. **Determine the required LN2 tank pressure.**

    We must know the LN2 storage tank pressure in PSIG. For our example, we will use a vertical tank with a *constant* pressure of 20 PSIG to overcome the total hydraulic friction.

3. **Do we know the temperature of the exhaust gases?**

Assume that we control the tunnel exhaust so that we get an exhaust temperature of about 20 degrees colder than the product end temperature and that we are using a spray tunnel. (Exhaust plenum only at the entrance of the tunnel) Hence, if we are freezing a raw product to an equilibrated temperature of -10°F, the exhaust gas temperature should be about -30°F. (See "Spray Counter flow" tunnels vs. "Convection" tunnels)

4. **Determine the optimum belt loading in pounds per square foot.**

5. **Determine the dwell time in minutes.**

6. **Determine the required production rate in pounds per hour.**

We can begin our calculations to freeze, for example, fillet of cod from 40°F to -10°F after having accomplished the following tasks:

5.1 Determine The Exact Nature Of The Product

• First, we establish if the product is unwrapped or wrapped, and if so what kind of wrapping. Let's for now assume the product is un-wrapped and that we are going to freeze raw cod fillets from 40°F to an equilibrated end temperature of -10°F. The physical dimensions are 3" wide x 4" long x ½" thick. The moisture content is known, and the standard data sheets show that cod has a moisture contend of 78%, an average freezing point of 28° F, and specific heat above freezing of 0.82 BTU/LB/°F, below freezing of 0.43 BTU/LB/°F and a latent heat of 112 BTU/LB.
If the product is wrapped with a plastic film of some sort, we must determine the dwell time by an actual test in a tunnel and record the temperature setting that freezes the product to its final desired temperature. The dwell time may be 2 to 3 times longer than an unwrapped product.

• With a raw product we can use physical characteristics found in the data sheets for cod. If we are dealing with a cooked cod, the physical product characteristics have changed and calculating the required energy can only be established by performing a calorimetric test procedure.

• To freeze the product to a temperature below its freezing point, we calculate the total required energy by completing the three steps as shown in Chapter 4.3, Example C. To chill the product indicates a temperature not surpassing its freezing point, in other words, not going beyond the latent heat of the product.

5.2 Determine the LN2 tank pressure

This is an important parameter that must be determined during these preliminary design steps. We must select a tank pressure that assures a constant flow of LN2 through the spray header during production. In addition, the LN2 tank pressure is necessary to calculate the conduit size between the tank and the tunnel. It must be sized so that the flow of LN2 to the freezer is constant at any tank level. A full tank exerts a higher pressure into the piping system than a tank that is only one third full. A *"Constant Pressure Control" should be included with the LN2 tank.* A supply of LN2 with a constant pressure entering the freezer's manifold is necessary to know that the product will be uniformly and consistently sprayed with LN2.

The average freezing system is designed with a vertical supply tank and an average pressure of 20 PSIG to overcome the line friction and head pressure requirements. Refer to Table 2 for more properties at different tank pressures.

5.3   Do we know the temperature of the exhaust gases?

We are using a spray-counter-flow tunnel for our model. The product is to be frozen to an equilibrated final temperature of -10° F. We can control the exhaust blower with a frequency speed control or with dampers in the exhaust ducting (not recommended), to get an exhaust gas temperature of about -30° F.

NOTE: This temperature is later used to calculate the energy available from the LN2.

5.4   Determine the belt loading

The belt loading must be accurately determined to size the freezer properly and is essentially what determines the capacity of the freezer. The belt loading is easily determined by laying out the individual product one piece at a time on a measured square foot. Each piece is 3" wide x 4" long and is arranged within a square foot with a minimal space between the individual pieces, such as 1/8".

Each piece weighs 0.25 LB. In each square foot we are able to layout 12 pieces x 0.25 LB = 3 LBS.

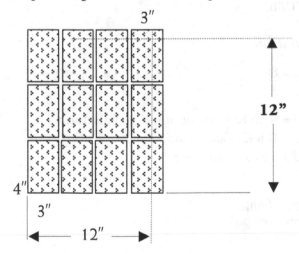

As we see, not exactly all pieces fit within a square foot when we include the spacing between the individual pieces. In addition, placing the pieces on the belt is usually accomplished by manual means and will never be as precise as on the illustration. Hence, we take a value of about 80% of the theoretical number we get from a layout as shown, which gives us a loading of 2.40 LB/FT^2.

5.5   Determine the dwell-time or freezing-time

The most common and accurate method to determine the dwell time is by actual freezing tests in a tunnel freezer with the same characteristics as the tunnel, selected to freeze a specific product. The product dwell time *is* necessary to determine the length of the tunnel freezer. Alternatively, the dwell time can be calculated using Plank's equation. The application of Plank's equation is complex and requires data of the food item, which is not always available. Such as (k) the thermal conductivity (BTU/hr/ft/°F), (h) the surface heat transfer coefficient (BTU/hr/ft²/°F) and (D) the density of the product. "Frozen Food Technology" edited by C.P. Mallett provides some insight in the complexities of the freezing process and the freezing-time of a product.

5.5.1   Calculating the Dwell-Time

The following is a simplified equation derived from empirical data that can be used to calculate the dwell-time with some accuracy:

From testing we find that it takes 8 minutes to freeze the meat patties from 40° F to (-10° F) in a "Spray Counter Flow" tunnel. Let's compare this by using the following formula:

$$T = (\Delta H) \times (W) \times (60) / (U) \times (Ap) \times (\Delta TLM) \text{ or } T = (\Delta H) (W) (60) / Q$$

Where:
T = Dwell time in minutes
Q = Individual product heat transfer rate in BTU/hr. = (U) (Ap) (ΔTLM)
Ap = Individual overall product surface area in ft² exposed to the refrigerant, i.e. thin patties, 2(.785 x D²) + (h x 3.146 x D). Sphere, (4 π x r²), etc.
ΔH = Specific enthalpy of the product (BTU/lb.)
W = Weight, each piece (lbs.)
U = Overall heat transfer coefficient, BTU/hr/°F/ft².

| For spiral freezers: | Product exposed | U = 8 |
| | Product wrapped | U = 3 |

For tunnels:	Product exposed	U = 10 to 12 (Consult tunnel manufacturer)
	Product exposed	U = 11 to12, Convection.
	Product wrapped	U = 4 to 5, Convection.

ΔTLM = Log mean temperature difference, °F

$$\ln \frac{|(T1 - \text{Freezer Temp.}) - (T2 - \text{Freezer Temp.})|}{|(T1 - \text{Freezer Temp.}) / (T2 - \text{Freezer Temp.})|}$$

Example 1: Beef patties, 5" diameter x ½" thick, T1 = 35°F, T2 = 0°F,

ΔH = 90% x 116.22 BTU's/lb. = 104.58 BTU/Lb.

U = 10, W = 0.25 lb. Freezer Temperature = - 80°F.

Ap = 2(0.785 x .42) + (3.146 x .42).042 = 0.715 ft².

$$\Delta TLM = \frac{\{35 - (-80)\} - \{0 - (-80)\}}{\ln \left| \frac{35 - (-80)}{0 - (-80)} \right|} = \frac{115 - 80}{1.44} = 24.3 \; °F.$$

Q = 10 x 0.25 x 24.3 = 173.78 BTU/hr.

T (min.) = (104.58) (0.25)x 60 / 173.78 = 9 minutes.

Example 2: Calculate how long it takes to freeze a raw tortilla dough ball in a LN2 spray tunnel with the following parameters, if only its moisture content is known:

1.   Size, 2" in diameter

2.   Weight, 45 grams or 0.0993 Lbs.

3.   Moisture content, 33.2 %

4.   Product freezing point, 32°F

5.   Initial temperature, T1 = 75°F

6.   Desired temperature, T2 = 28°F

7.   LN2 tank pressure, 20 psig

8.   Tunnel set point temperature. -160°F

9.   Type of tunnel, Spray Counter flow

First determine the product's specific heats and the latent heat of fusion.

These equations are given in paragraphs 4.4.0 and 4.4.2

1.   To calculate the specific heat *above* freezing we use:
       (% Water)/100 + .42(1 - %water/100)

2.   To calculate the specific heat *below* freezing we use:
       0.5(% water/100) + .42(1 - % water/100)

3.   To calculate the latent heat we can use: 144 (% water/100)

Now we can execute the calculations:

1.   Sp.Ht. AF = 33.2/100 + 0.42(1 – 33.2/100) =0.613 BTU/Lb.

2.   Sp.Ht. BF = 0.5(33.2/100) + 0.42(1 – 33.2/100) = 0.307

3.   LH of fusion = 144(33.2/100) = 47.808

Plank's equation requires the value of the energy to freeze the product (ΔH) to its desired end temperature to be expressed in BTU/lb.

We can use Chaim's equation to find ΔH: $H_{Total} = H_A + L + H_B$

H_{Total} = 0.613(75°F – 32°F)+(.90 x 47.808) + 0.307(32°F – 28°F) = 70.614 BTU/lb.

Now we can continue to calculate the dwell time using Plank's equation, paragraph 5.5.1:

$$T = (\Delta H) (W) (60) / Q$$

Q = Individual heat transfer rate in BTU/Hr = (U) (Ap) (ΔTm)

U = 10 BTU/hr/°F/ft².

Ap = $(4\pi \times r^2)$ = 4 x 13.4 x .0069 ft².   = 0.375 ft²

ΔTLM = $\dfrac{\{(75° - (-160°)\} - \{(28° - (-160°)\}}{\ln \left|\dfrac{75° - (-160°)}{28° -(-160°)}\right|}$ = 39.17° F

Now, T = $\dfrac{(70.614 \text{ BTU/lb}) (0.0993 \text{ Lb.}) (60)}{10 \times 0.375 \times 39.17}$ = 2.90 minutes (3 minutes)

NOTE: The value of the latent heat of the product was multiplied by (0.90) to signify that 10% of the moisture remains un-frozen.

The Thermal Transfer or Heat Transfer Coefficient (U) of a freezer is a function of the physical properties of the LN2 used at the time and the internal air velocities. In addition, the type of freezer, convection or spray tunnel and the manufacturer of the freezer determine the value of (U). It is recommended that the manufacturer of the freezer is consulted.

◊ Equilibrated temperatures are reached after the product has some time to equalize the surface temperature, which is much lower in temperature than the final temperature, with the core temperature of the product.

◊ Do not measure the final product temperature directly after it exits a cryogenic freezer. Wait about 10 minutes for a tunnel freezer and about 5 minutes for a spiral freezer. This will be the final equilibrated temperature of the product.

5.6   Determine the production rate

The production rate is generally the known factor in the equation. We will use 1000 Lbs/HR for our example calculation.

Lbs/hr. = $\dfrac{\text{Belt Loading x Freezing Zone x 60}}{\text{Dwell Time(Minutes)}}$

Where:
The freezer's production rate is in Lbs/hr.
Belt Loading is the weight of the product on one linear foot of belt.
Freezing Zone is the length of the freezer minus the inlet and outlet extensions in feet.
Dwell Time is in minutes.

5.6.1   *Theoretical vs. Practical Production Rate*
It is important to distinguish between the terms "Theoretical Production Rate" and the "Actual

Production rate". The theoretical production rate refers to a condition where the product exits the freezer at the target temperature with the maximum belt loading, at a steady production feed during the entire work day. In the practical world, product in feed into the freezer is influenced by stoppages due to conditions up-stream or down-stream from the freezer, variations in the product loading density on the conveyor and often higher initial product temperatures.

These issues must be taken into account when the freezer's capacity is being evaluated.

5.7.1   The Conveyor Belt
Introduction
Typical liquid nitrogen tunnels are equipped with Stainless Steel USDA approved conveyor belting. The standard belt is the B18-16-16 with openings of 9/16" (See Fig.9 full size sample) This size belt is selected to maximize the circulation effect in the tunnel and to minimize the mass and hence the heat load of the tunnel. The belt has been installed with several master links designed to permit separation of the belt for tension adjustment and maintenance. After some period of operation, a conveyor belt will stretch and requires adjustment of the tension. Often, a section of the belt needs to be removed to correct the belt tension.

The conveyor belt can be equipped with product guide rails of 1" to 4" high to keep product from being blown off the belt. The selection of these guide rails will be determined during the initial design stages.

The belt should have the proper tension and scheduled inspections are recommended.

NOTE: A detailed instruction manual is supplied with a new tunnel. Follow these instructions diligently or the freezer can incur serious damage.

Lubrication
Always use low temperature food grade grease as recommended by the manufacturer. Tunnels require very little lubrication. Tunnels and other cryogenic freezers manufactured prior to 1990 may require lubrication. Each shaft has two bearings; each bearing has one grease nipple and should receive grease until it oozes out around the races. Inspection of the bearings will determine the frequency of lubrication. More up to date freezers are manufactured with bearings made from a self-lubricating material, called UHMW Teflon.

A low temperature food grade lubricant is:

"TRIBOLUBE – 21" from Aerospace Lubricants, Inc.

(614) 291-3045 Fax. (614) 291-7416

5.7.2   Belt Loading
The capacity of the tunnel has been calculated with a very specific belt loading. The formula used to determine the belt loading is:

Tunnel Capacity (Lbs./hr.) = $\dfrac{\text{Usable belt length x Belt Loading x 60}}{\text{Dwell Time}}$

Tunnel Capacity =   Required production in Lbs. /hr.

Usable belt length = The length of the belt within the tunnels insulated shell.

Belt Loading =   The amount of product that can be arranged on the belt within a 12" long area of the belt in lbs. The usable belt width is the overall belt width minus 2".

Dwell Time =   The time in minutes required to freeze the product.

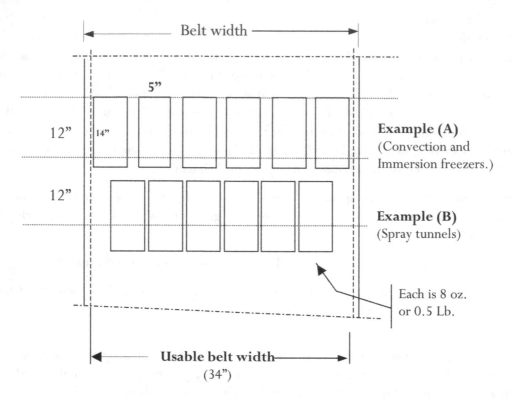

Example (A)
(Convection and Immersion freezers.)

Example (B)
(Spray tunnels)

Each is 8 oz. or 0.5 Lb.

5.7.3   *How to Arrange the Product*

1.   **Spray Counter flow Tunnel**

     The belt loading should be arranged to cover the belt area as efficiently as possible. We have learned that the product can be arranged on the belt with very little to no space between the individual pieces (1/8" is enough) such as shown in Example (B):
     Belt loading per Linear Foot = (12/14 x 0.5 lb.)x 6 pieces = 2.57 lbs.

2.   **Convection Tunnel**

     With a convection tunnel it is recommended to have better circulation around the product for a more efficient heat transfer. The spacing between the pieces should be around 1/2". See Example A.

     For the belt loading calculations for product consisting of small pieces such as shrimp or meat balls, the best method is to arrange and record the product weight on a square foot area with a density of 80% for a convection tunnel and 99% for a spray counter flow tunnel. The pieces should not be layered.

3. **The Triple Deck Tunnel**

No matter which spray method has been selected in a triple deck tunnel, always freeze the product on the top belt with a crust deep enough so that no sticking or deformation will take place when the drop to the next belt is made.

Small products such as shrimp, diced meats etc. can be multi layered on the next two belts during the equilibration cycle. This is accomplished by slowing down the two lower belts. Make sure that the tunnels have been fitted with product guide rails high enough to contain the product on the belt.

For the standard triple deck tunnel the clearance between the belts determines how *long* the product can be without deforming or jamming when transferring from belt to belt. Consult your equipment manufacturer.

4. **The Immersion Tunnel**

When the immersion freezer is equipped with a product feed system which emulates a straight tunnel, in-feed on one side and exit on the opposite side of the freezer, the belt loading should be exactly as recommended as for the convection tunnel. When the immersion freezer is fitted with a top-loading feature, product feed methods should always be at such a rate as to avoid the product from sticking to the belt when dropping into the LN2 bath. Adjusting the belt speed, LN2 bath depth and the product feed rate controls this.

*For breaded products, consult a specialist.

IMPORTANT: The LN2 level should be adjusted to a height precisely even with the product height or slightly lower. Never completely submerge the product in the LN2. This will only increase the consumption rate.

Fig.8   CONVEYOR MESH BELT
B84-16-17 – (FULL SIZE)

84 = Weaves per foot of belt width
16 = Rods per foot of belt length
17 = Gauge of the wire overlay.

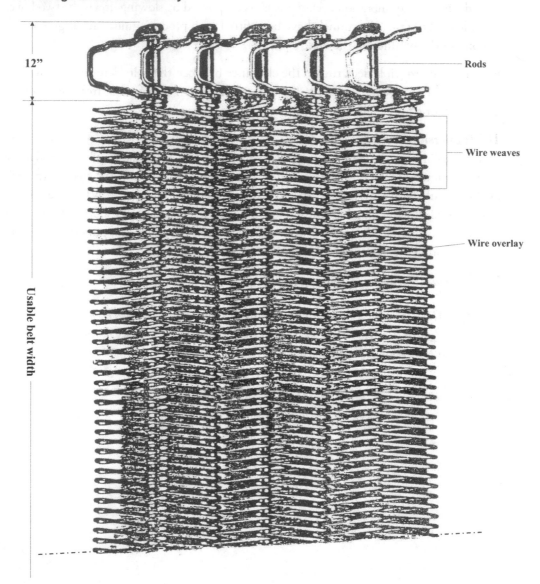

12"

Rods

Wire weaves

Wire overlay

Usable belt width

Fig.9   CONVEYOR MESH BELT
B18-16-16

18 = Weaves per foot of belt width
16 = Rods per foot of belt length
16 = Gauge of the wire overlay

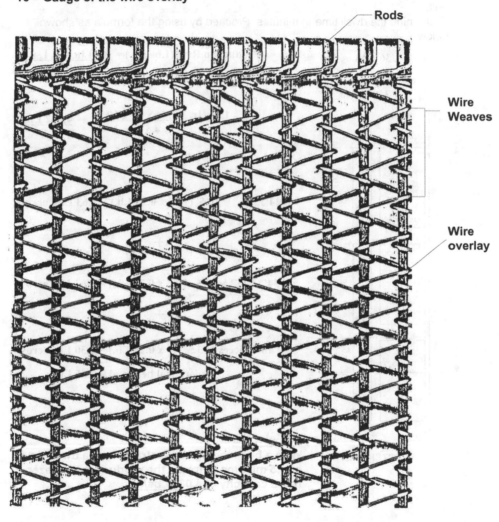

Rods

Wire
Weaves

Wire
overlay

5.8 FREEZER CAPACITY EVALUATION
(Sample Calculation)

Task: Determine the production rate of a LN2 freezer with a 36" belt, and a freezing zone of 40'.

First determine the belt loading in pounds by placing the product on a linear foot of belt.
Determine the dwell time in minutes. Proceed by using the formula as shown below.

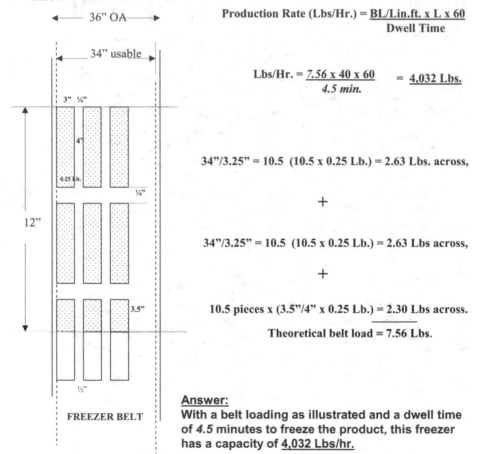

Production Rate (Lbs/Hr.) = $\dfrac{BL/Lin.ft. \times L \times 60}{Dwell\ Time}$

Lbs/Hr. = $\dfrac{7.56 \times 40 \times 60}{4.5\ min.}$ = **4,032 Lbs.**

34"/3.25" = 10.5 (10.5 x 0.25 Lb.) = 2.63 Lbs. across,

+

34"/3.25" = 10.5 (10.5 x 0.25 Lb.) = 2.63 Lbs across,

+

10.5 pieces x (3.5"/4" x 0.25 Lb.) = 2.30 Lbs across.

Theoretical belt load = 7.56 Lbs.

Answer:
With a belt loading as illustrated and a dwell time of *4.5* minutes to freeze the product, this freezer has a capacity of <u>4,032 Lbs/hr.</u>

SECTION B

HOW TO OPERATE
LN2 FREEZERS PROPERLY

6. PROCESS AND OPERATIONAL CONSIDERATIONS

6.1   Product refrigeration load

6.2   Cool down requirement

6.3   Fixed operating load

The operators of cryogenic freezers should understand not only how the freezer functions, but which specific adjustments to the freezer will determine the LN2 consumption. It is not one area of the freezer's operation that determines how much LN2 is used for a specific freezing task but a series of operational adjustments that will ultimately influence the consumption rate.

The LN2 consumption rate is determined by the sum of the three following factors:

6.1   Product refrigeration load

This is the LN2 required for cooling or freezing the product, $(\Delta H/Q_{Total})$.

Where ΔH is the total product enthalpy to cool or freeze from T1 to T2, and

Q_{Total} is the LN2 requirement to remove the product heat load. Refer to Table 2 for the base value of Q_{Total}

The product refrigeration load is the largest heat load in any freezing system.

COMMENT: Not much the operator can do here.

6.2   Cool down requirement

This is the amount of energy (LN2) required to cool down the freezer to the desired freezing temperature or set point temperature. Good management of this procedure can save the LN2 consumption at the end of the day. During an eight hour production run, the freezer is going through a cool down procedure at least 3 times. The first time will be at the morning start up. The second and third times are less energy intensive and are generally required after the crew returns from brakes.

COMMENT: The first morning start up should begin not more than 30 minutes before the first product pieces enter the freezer. Contact the equipment manufacturer if it takes more than 30 minutes to cool down the tunnel.

Don't leave the freezer running with the minimum set point temperature during breaks! Increase this temperature to a temperature close enough to the set point temperature so that the required set point temperature can be reached within 5 minutes.

Do not forget to reduce the exhaust blower suction when increasing the temperature of the tunnel during breaks.

6.3   Fixed operating load

This is the amount of LN2 required to maintain the freezer at the set point temperature to overcome the heat input from the circulating fans and the heat leak through the insulated walls, etc. Here we assume that the exhaust system is adjusted properly. This is in reference to the 20°F deferential for a spray tunnel exhaust vapors and the product equilibrated end temperature. Thus a tunnel processing a product which requires an equilibrated end temperature of 10°F should have an exhaust temperature of approximately -10°F.

The values of paragraphs 6.1 and 6.2 are determined by a simple calculation. The value of paragraph 6.3 is more difficult to verify. It is mathematically determined by adding the heat inputs of the circulating fans, the freezer walls and the extracted exhaust vapors.

This method is seldom used due to the unpredictable exhaust losses. One operator may adjust the amount of exhaust differently than the other. Actually, it is simpler to add the LN2 consumption within a specific time period (30 days for example) and divide that number by the product frozen within that same period.

For example, in precisely 30 days 195,671 LBS of fillets were processed. Within the same 30 days we used 200,000 LBS of LN2. This means that we have a system freezing the product at a rate of 200,000 divided by 195,671 = 1.022 LB/LB. The calculated number is 0.88 Lb/Lb. (Example C, in Chapter 4.3) we used 1.022 Lb/Lb. We can now safely conclude that this freezer has an efficiency of 86%.

> *CAUTION: Operators not trained properly should not be allowed to make any adjustments to the freezer controls. As a matter of fact, they should not operate the freezer at all for two important reasons. Safety considerations and ultimately the freezing cost.*

7. KEEPING AN EFFICIENT LN2 FREEZING OPERATION

7.1 Critical Guidelines to follow

7.1.1   *The LN2 Tank Must Be Filled with a Minimum of Vapor Loss*

7.1.2   *Keep the Tank Pressure Constant at any LN2 Level and During the Filling Process*

7.1.3   *Design the Piping System to Allow a System Pressure of not more than;*

7.1.4   *The Piping System, including the Freezer's LN2 Spray Headers, should have a Pressure Drop of not more than 3 Psig*

7.1.5   *The Exhaust Vapors should have a Temperature as Close as Possible to the Product Equilibrated Temperature*

7.1.6   *Constantly Monitor the Vapor Balance at the Exit and Entrance of the Tunnel*

7.1.7   *With the Freezer Operating at its Designed Freezing Capacity, the Set-point Temperature Should Not Exceed its Designed Operating Temperature*

7.1.8   *Maintain the Maximum Belt Loading*

7.1.9   *During Long Production Intervals Raise the Set-point Temperature and Reduce the Exhaust Draw*

7.1.10 *Make Sure that all Circulating Fans are Running*

7.1.11 *Make Sure that all the Injection Nozzles are Not Plugged*

Figure 10   Typical Cryogenic Freezing System

7.1   Critical Guidelines to follow

7.1.1   *The LN2 Tank Must Be Filled with a Minimum of Vapor Loss*

Use the "TOP FILL" valve as much as possible

Avoid pumping at 100% of the pump capacity

7.1.2   *Keep the Tank Pressure Constant at any LN2 Level and During the Filling Process*

Use a Pressure Switch regulated tank pressure control

7.1.3   *Design the Piping System to Allow a System Pressure of not more than;*

20 psig for vertical tanks with a total pressure drop of +/- 3 psig.

5 to 10 psig for horizontal tanks with a total pressure drop of < 1psig.

7.1.4   *The Piping System, including the Freezer's LN2 Spray Headers, should have a Pressure Drop of not more than 3 Psig*

7.1.5   *The Exhaust Vapors should have a Temperature as Close as Possible to the Product Equilibrated Temperature*

7.1.6   *Constantly Monitor the Vapor Balance at the Exit and Entrance of the Tunnel*

7.1.7   *With the Freezer Operating at its Designed Freezing Capacity, the Set-point Temperature Should Not Exceed its Designed Operating Temperature*

This Temperature is the Average Tunnel Temperature and is Generally -160°f

7.1.8   *Maintain the Maximum Belt Loading*

7.1.9   *During Long Production Intervals Raise the Set-point Temperature and Reduce the Exhaust Draw*

Use a PLC controller. This will reduce the freezing cost.

7.1.10   *Make Sure that all Circulating Fans are Running*

The Tunnel Heat Transfer Coefficient (U) depends on it and will influence the Heat Transfer Rate (W)

7.1.11   *Make Sure that all the Injection Nozzles are Not Plugged.*

Inspect this at least twice a year. Install an accurate pressure gauge just before the header. When the pressure creeps up, the nozzles are beginning to clog.

Fig. 10   TYPICAL CRYOGENIC FREEZING SYSTEM

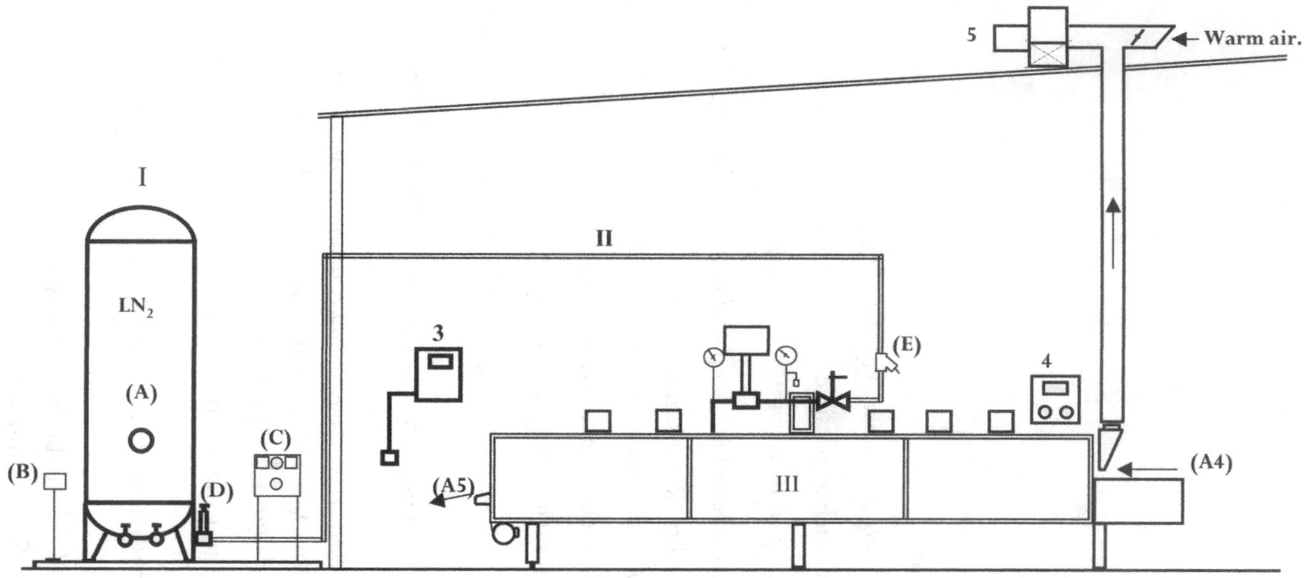

SYSTEM COMPONENTS

I. Select the most efficient LN2 supply system:
   (A).. Low pres.. super insulated receiver.
   (B).. "Telemetry" Liquid Level monitor.
   (C).. Constant Pressure Regulator.
   (D).. Vacuum Insulated take–off valve.
   (E).. Strainer with blow down.
II. LN2 Supply piping. Vacuum or PU insulated.
   3. Oxygen Monitor.

III. Follow recommendations for tunnel selection
   (A4).. Product infeed. Do not overload !
   (A5).. Transfer frozen product to blast freezer ASAP.
   4. Temperature Control center. PLC (Local or remote)
   5. Exhaust blower and controls.

8.   CRITICAL COMPONENTS OF A LN2 FREEZING SYSTEM

8.1   LIQUID NITROGEN STORAGE TANKS
8.1.1   The LN2 Tanks
8.1.2   The Vacuum
8.1.3   The LN2 Tank Size
8.1.4   The LN2 Discharge Connection
8.1.5   Maintaining a Constant Pressure in the LN2 Tank
8.1.6   Filling the LN2 Tank Properly
8.1.7   LN2 Storage Losses

8.2   LN2 Transfer Piping
8.2.1   Polyurethane Foam Insulated Piping
8.2.2   Super-Insulated Vacuum Jacketed Piping

8.3   Behavior of cryogenic fluids in pipelines
8.3.1   The Geyser Effect
8.3.2   Economic Evaluation of VJ vs. PU Foam Insulated Pipe
8.3.3   When to Select VJ or PU Insulated Pipe
8.4.0   Heat Leak into a (LN2) Transfer Line
8.4.1   Calculating the LN2 Pipe Size
8.4.2   Flow Reduction Due to a Dual Phase Condition

8.5   TYPE OF FREEZERS
8.5.1   The "Spray Counterflow" Tunnels
8.5.2   The Process
8.5.3   The Convection Tunnel
8.5.4   The Tri-Deck or Triple Pass Tunnels
8.5.5   The Spiral Freezer
8.5.6   Belt Types
8.5.7   Product Clearance
8.5.8   Circulating Fans
8.5.9   Night Injection Feature
8.5.10   Access Doors
8.5.11   The Batch or Cabinet Freezer

8.1 LIQUID NITROGEN STORAGE TANKS

The liquid nitrogen storage tank is the most important component of the freezing system. Without the correct tank design and management, the freezing installation will be dysfunctional. The topics described in this chapter are extremely important for the total efficiency of the freezer and the plant engineer(s) should be familiar with each topic or they may eventually be faced with unexplainable high freezing cost.

8.1.1 The LN2 Tanks

The cryogenic tank manufacturers offer vertical and horizontal storage tanks. The vertical tanks come in sizes ranging from 500 to 15,000 gallons with standard operating pressure ratings from 75 to 250 psig. The tanks used with low pressures as is necessary with food freezers are the 75-psig pressure rating category. The safety devices on the tank are factory installed and are set to release at the maximum working pressure of the tank as indicated on the ASME plate attached to each tank. The actual maintained tank pressure which is required to push the LN2 through the piping system to the freezer is of course much lower, 10 to 20 psig. These matters are not always taken care of by the industrial gas company contracted to deliver the LN2. It is therefore necessary to be aware of these issues and it *must* be part of a training routine that the gas supplier performs for the plant staff before the process is put into service.

Horizontal tanks come in sizes ranging from 1000 to 50,000 gallons with similar pressure ratings and pressure relief protections as the vertical tanks. Horizontal tanks in cryogenic service are often placed on pedestals 10 to 30 feet high to allow the LN2 to flow to the freezer by means of natural gravity. It is not uncommon to have a system operating with a tank pressure of 5 psig to get the maximum BTU's from the LN2.

REMINDER:The Caloric value (BTU) of LN2 in a freezer is reduced when the tank pressure is increased as shown in Table 2. Maintain as low as possible tank pressure!

Be prepared, you will be reading this at least another 10 times, it's that important.

8.1.2   The Vacuum

When the LN2 tank is leased from the gas supplier, the vacuum measurement and maintenance of the LN2 tanks are not procedures plant personnel are responsible for but which they must understand and monitor. If the annular space between the outer tank shell and the inner vessel is occupied with a high quality material (such as layers of aluminum foil and aluminized Mylar film, which have been wrapped around the inner vessel) the tank is specified as a "Super Insulated" tank. These tanks have a normal evaporation rate of 0.1%/day of its content. If the annular space is packed with Perlite, the normal evaporation rate is 0.25%/day of its content. A vacuum is then pulled to levels lower than 200 microns, measured when the tank is warm. When the vacuum is measured during a routine maintenance check, the tank is cold and contains LN2; the vacuum level should be at 20 microns or lower. If the LN2 tank is rented, insist that your gas supplier provide you with a copy of these readings and if there is any doubt of the correctness of a good vacuum you should insist on being shown the gauge readings during the pumping process after the tank has been secured on a concrete pad at the food processing plant.

The vacuum in the tank's annular space should be checked once every 5 years for a new tank and yearly for a tank older than 10 years.

A poor vacuum will result in a high freezing cost. Use super insulated tanks only for low pressure applications.

8.1.3   The LN2 Tank Size

Why should we be concerned with the size of the storage tank? Again, hopefully we have not forgotten that the LN2 storage tank plays a pivotal roll in an efficient freezing system. It should have a good vacuum and be sized precisely right for the amount of nitrogen the freezer will use per week.

◊ For a freezing tunnel installation, the LN2 tank should be refilled not less than once a week or once every two weeks when the process runs 6 days per week. Only then can we be sure that we are using a super-critical liquid free of vapor.

Since the contents of the tank is a cryogenic fluid which has a temperature of (-320°F) at one atmosphere (0 psig), we must be cognizant of the fact that with time, the LN2 in the tank will absorb heat and begins to increase in pressure and will be saturated with vapor. This nitrogen vapor is not desirable in the freezing tunnel; it will reduce its efficiency. Moreover, the tank will eventually release this vapor to maintain a pre-set pressure in the tank. Venting of the nitrogen vapors must be kept to a very minimum if you want to have an economic freezing system. A storage tank which vents often must be avoided at all cost. Costly vapor vent losses can be avoided by selecting the correct tank size in relation to the LN2 consumption of the freezer(s). A delivery of "still" super cooled LN2 once a week will maintain a super-critical liquid in the tank, assuming that the delivery tanker arrives with a low pressure, 10 to 20 psig, and is *not* venting.

◊ It is advisable to know where the vent valves are located on the tank. Inspect these devices often. If it releases or vents nitrogen on a regular basis the gas supplier must be consulted.

If possible, when required, the storage tank should be replenished late on Saturday if Friday is the last and Monday will be the next working day. Of course, Monday morning before the start of the production run is ideal.

Example: The LN2 usage of the freezer is 1000 pounds per hour (1000 Lbs/hr.), 8 hrs/day and 5 days/week. The consumption per week will be 1000 x 8 x 5 = 40,000 Lbs/week or 5,924 Gal./wk.

A reserve of at least 2 days of production should be kept in the tank to safeguard against delivery delays.

In 2 days we could consume (1000 x 8) x 2 = 16,000 Lbs. or 2,370 Gallons.

The total amount of nitrogen we have available per week is then 5,924 Gal. + 2,370 = 8,294 Gals.

The standard size for this usage is a 9,000 gallon tank.

Understandably, the cryogen supplier prefers to unload the entire contents of one delivery trailer into the storage tank to reduce the distribution cost. A LN2 tanker has a net content of 6,444 gallons at an average pressure of 10 psig. In the example, at the end of a 5-day week, 2,370 gallons remains in the storage tank (The reserve). With the addition of the 6,444 gallons of the delivery tanker completely transferred into the 9,000 gallon LN2 storage tank we now have a total amount of 8,814 gallons of low pressure super-critical LN2 in the tank. Discuss the correct amount of storage with the gas supplier. At times, for the extreme climate regions of our country, when the food processing plant is more than 100 miles from the air separation plant, some extra reserve is recommended.

8.1.4   *The LN2 Discharge Connection*
Vertical liquid nitrogen tanks are fitted with numerous take-off valves. Each plays an important function in keeping the liquid nitrogen readily available for use as a liquid or vapor. Freezing systems require liquid nitrogen (LN2) to be piped to the tunnel's control manifold and spray header. During the freezing operation, the LN2 is sprayed into the tunnel by means of a modulating control valve, which requires a constant flow and pressure to function properly. LN2 tanks have two take-off valves that can be used for liquid withdrawal. One is labeled "AUXILIARY LIQUID" and one is labeled "LIQUID TO ECONOMIZER". Only use the "AUXILIARY LIQUID" withdrawal valve. The liquid to economizer withdrawal is connected to a siphon pipe inside the inner vessel to be used for vapor requirements. *DO NOT use this connection for any freezer.* It will stop flowing LN2 at levels around 30% of the tanks capacity.

Horizontal tanks have a much simpler piping scheme. This type of tank is generally used to supply *liquid* to a specific application and is therefore capable of maintaining a very low pressure. There are actually only three main lines available. A fill line, a liquid withdrawal line, and a vent line. Due to the very large cold liquid surface area, pressures in these tanks are very stable and often do not require the constant pressure regulating device always necessary with the vertical tanks.

A good supply of LN2 into the main conduit to the freezer is assured by using a vacuum insulated liquid withdrawal and valve of appropriate size. Request this type of valve from your gas supplier when the freezer(s) produce in excess of 2,000 Lbs/hr.

8.1.5   *Maintaining a Constant Pressure in the LN2 Tank*
The maintenance of a *constant pressure* in the LN2 tank is always overlooked when designing a cryogenic freezing system, although it will influence the efficiency of the freezer. A constant

pressure and a constant flow of LN2 to the freezer go hand in hand and must be one of the first design features considered when the engineers are designing the freezing process. First, make sure the LN2 line to the freezer is connected to the proper valve on the tank. Secondly, select a constant pressure device large enough to maintain the tank pressure constant for the designed maximum instantaneous flow rate of LN2 to the freezer. At all times should this device be able to maintain a constant pressure (within one PSI) in the tank and hence in the pipeline to the freezer. This includes conditions at any liquid level in the tank, minimum or maximum level, *and* during the filling process.

8.1.6   *Filling the LN2 Tank Properly*

> *Excessive bleed-off or vent losses will occur when the filling process is done incorrectly. This contributes to a freezing cost increase.*

The food processor is never involved with the "How to" details of filling the liquid tank i.e. transferring liquid nitrogen from the tanker to the storage tank. However, it is a *major* contributor to a high freezing cost if not done properly. The transfer of LN2 from the delivery tanker to the storage tank is performed by a centrifugal pump connected to the fill connection on the tank by means of a flexible stainless steel hose. The transfer pump can operate between 10% and full speed. The speed of the pump should be adjusted so that the liquid nitrogen flows into the storage tank with a velocity low enough as not to increase the storage tank original pressure. In theory, this is of course impossible. The hydraulic pressure caused by the rising liquid level must go somewhere. The tank manufacturer has therefore installed two fill valves. One "TOP FILL" valve and one "BOTTOM FILL" valve. As the names imply, the top fill valve will send LN2 to the top of the tank and the bottom fill valve will send the LN2 only to the bottom of the tank.

Liquid nitrogen is a very unstable cryogenic liquid when pumped and will enter the tank as a vapor saturated liquid. This vapor will ultimately overcome the vapor pressure above the liquid and accumulates in the free space above the liquid increasing the pressure in the tank. If this cycle is not kept in check, the tank pressure will rise and vapor is released through a safety valve or the pressure control device.

The pressure in the storage tank is kept close to constant during the filling process as follows:

1.   First, open the "Bottom Fill" valve about 30% and the "Top Fill" valve to almost 100%. Most likely, the experience of the operator will determine how the two valves are manipulated.

2.   During start-up do not run the pump at 100% of pumping capacity. Begin at 50% or less and observe how the tank pressure reacts when the pump is started.

3.   Keep pumping with these initial settings for about 5 minutes. Observe the storage tank pressure. If the pressure remains at the original setting, the operator can do two things. Increase the pumping rate or open the "Top Fill" valve somewhat more. Remember that a top fill entrance in the tank's headspace will drop the tank pressure by condensing the nitrogen vapors in the headspace of the tank.

4.   Continue pumping by controlling the top and bottom fill valves.

5.   Make sure that the tank is not venting off vapors at an elevated rate at any time during the filling process. This is wasted nitrogen and expensive for the customer.

6.   When the liquid level is at 60% full, the bottom fill valve should be closed and only the top fill valve is used to complete the pumping process.

Hydrodinamically, when a liquid is pumped into a tank, an increase in pressure is unavoidable. That is why the release of pressure during the filling of a nitrogen tank is unavoidable as well. So do not panic if you see this happen. However, when the filling process is completed and the main tank relieve valve and the pressure control panel are relieving pressure, the filling process was done to quickly or erroneously. Ideally, the liquid transfer to the storage tank is accomplished with some venting during the filling process and no venting after the filling process is completed.

8.1.7   LN2 Storage Losses
1.   If the daily LN2 consumption rate is lower than calculated, the pressure in the tank will rise and the pressure control device will release nitrogen vapor to maintain the tank pressure at the set pressure. The main safety valve is set at a much higher pressure than the running pressure to the freezer. Typical settings are as follows for a freezing operation:

   1. Main safety valve, 150 or 250 PSIG.

   2. Weep valve, 100 or 200 PSIG. (Diaphragm type regulator)

   3. Pressure control device, 20 PSIG for a vertical tank.

2.   If, on the other hand the daily consumption is as calculated, the tank pressure will drop and the pressure building device will maintain the exact tank pressure. No losses will occur.

NOTE: A storage tank with a good vacuum can remain idle (no production) without losing any LN2 for at least 3 days and 100°F ambient temperature.

FIGURE 11 LOADING AND UNLOADING OF LN2 TRANSPORTS
To operate a cryogenic freezing system effectively one should understand where the LN2 comes from before it arrives at the processing plant. The LN2 should arrive as a super critical liquid, should not be venting, and should have a pressure of < 20 psig.

The storage tanks at the air separation plant maintain a pressure between one and five psig and can hold as much as 69,000,000 cubic feet or 741,060 Gallons of liquid nitrogen. Approximately 25 Kilowatt of power is required to produce one cubic feet of Liquid nitrogen.

FIGURE 11 LOADING AND UNLOADING OF LN2 TRANSPORTS

AT THE AIR SEPARATION PLANT

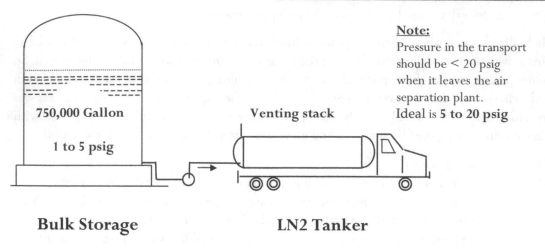

Note:
Pressure in the transport
should be < 20 psig
when it leaves the air
separation plant.
Ideal is **5 to 20 psig**

750,000 Gallon

Venting stack

1 to 5 psig

Bulk Storage

LN2 Tanker

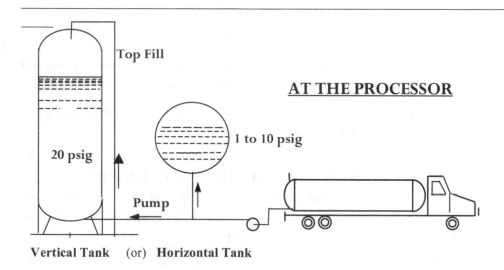

Top Fill

AT THE PROCESSOR

1 to 10 psig

20 psig

Pump

Vertical Tank (or) **Horizontal Tank**

The pressures indicated above are important to know since the pumping process to the storage tank will increase the tank pressure and may contribute to losses up to 2% of the transferred liquid. It is therefore essential, to keep the pressure in the liquid transport as low as possible and to pump at a low rate.

Fig. 12   Typical LN2 Tank Piping Scheme

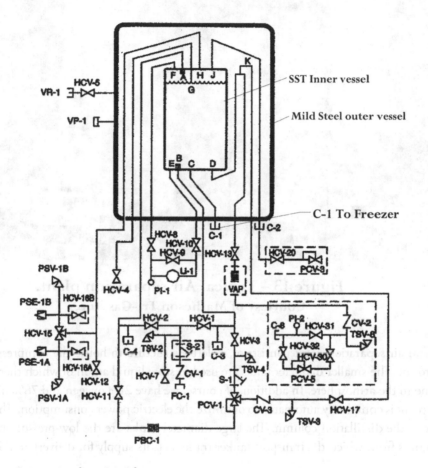

Valve C-1   Auxiliary Liquid to process.
Valve HCV-2 Top fill valve
Valve HCV-1 Bottom fill valve
Valve HCV-13 For vapor use only.

Figure 13 – Typical Air separation plant.
Courtesy of Matheson Tri-Gas, Inc.

The typical air separation plant consists of, at left a structure to house the compressors and the control room. The smaller diameter tanks are used to store liquid argon of which there is only 1% by volume in the atmosphere. In addition, of course, we have 21% oxygen and 78% nitrogen. The modern plant is completely automated to optimize the electric power consumption. The tall square structure is the distillation column. The large diameter tanks are the low-pressure nitrogen and oxygen tanks from which the transport tanker trailer gets its supply for delivery to the processor.

8.2   LN2 Transfer Piping

Always overlooked, but if not designed properly, the conduit or piping between the storage tank and the freezer will have a profound effect on the efficiency of the freezing system as a whole. Consult an engineer well informed in the design of LN2 piping systems for low pressure applications to assist you.

Two types of LN2 transfer conduit or pipe is available for industrial cryogenic applications. (1) Polyurethane foam insulated copper or stainless steel piping and (2) Super insulated vacuum-jacketed piping (VJP) This chapter describes the advantages and disadvantages of the two different piping systems. Inherently, cryogenic fluids such as LN2 present many challenges to the designer. For our purpose, specific attention should be given to (A) The tank pressure and subsequent pressure drop in the piping system and (B) the heat leak of the piping system between the cryogenic tank to the freezer. When close attention to these two fundamental issues is given, a good quality cryogen will be available to the process i.e. sub critical fluid with a minimum of vapor.

8.2.1   Polyurethane Foam Insulated Piping

In the practical world, the cost of the components necessary to complete the freezing installation is basically a starting point to "cut corners". One usually begins by selecting piping based on the initial cost such as polyurethane (PU) insulated piping. Yet, this conduit between the storage tank and the freezer is the second most important component of the installation. By selecting UP insulated piping, regardless the length, the propagation of a poor quality cryogen has begun. As a result, the heat gain has become a factor in propelling a two-phase (liquid & vapor) cryogen through the conduit towards the freezer. As we have learned in other chapters, large quantities of nitrogen vapors, more than 5%, released in the cryogenic spray zone of the freezer can reduce the freezing capacity of the freezer and as a result will increase the specific LN2 consumption.

As a practical matter of fact however, some vapor in the LN2 entering the freezer is unavoidable with the technology currently available. Two-phase LN2 flow is a complex phenomenon and mathematical models have been developed to understand the hydrodynamics and heat transfer in cryogenic piping systems causing this two-phase flow. Hence, the design of cryogenic piping will improve if one understands these theories. It is sufficient to say that any heat gain trough the pipe wall will disturb the liquid equilibrium of the cryogen within the pipe and should be avoided. If the *operating cost* is to be kept at a minimum, select vacuum jacketed pipe.

PU insulated conduit can be used when:

1.   The conduit has an *equivalent* length of less than 100 feet and,

2.   The flow of cryogen is more or less constant.

3.   The tank pressure is 20 psig or less

4.   The tank is on a higher level than the application equipment

In the publication "Cryogenic Two-Phase Flow" by N.N. Filina and J.G. Weisend II one can study this two-phase flow phenomenon in depth.

8.2.2   Super-Insulated Vacuum Jacketed Piping

This is a highly efficient piping system with a very low heat loss that will yield a lower operating cost over time vs. the polyurethane foam insulated piping. There are two types of VJP available from a variety of manufacturers. The earlier piping systems required a vacuum pump to be installed to maintain the vacuum in the annular space between the actual inner transfer pipe and the outer jacket. A more advanced static piping system is now becoming the norm in the industry. This VJ pipe is designed to perform for many years without degradation of its vacuum and is warranted for as long as 10 years. A vacuum of 9 microns or less is permanently sealed in the annular space between the inner and outer pipe. The piping is pre-fabricated by the manufacturer and is quickly installed between the LN2 tank and the freezer.

Use VJ pipe exclusively when the total equivalent-length of the piping system is more than 100 feet. Equivalent-length is the measurement of the pipe plus adding in linear feet make-up fittings such as elbows, tees and valves. Tables are available to get the equivalent lengths for these fittings.

VJ pipe transfers the LN2 as a super cooled liquid from the storage tank to the freezer. A pressure differential causes the LN2 to flow to the freezer with a minimum of heat leak from the surrounding atmosphere.

Average Heat loss values: For VJ pipes, 0.45 BTU/ft/hr.

For PU pipe, 20 BTU/ft/hr. New

For PU pipe, 200 BTU/ft/hr. > than 5 years old

8.3   Behavior of cryogenic fluids in pipelines

When traveling through a pipe from point (A) to point (B) cryogenic fluids behave differently than water. To a large extend this often unpredictable behavior is contributed to the large temperature differential (ΔT) between the cryogen and the ambient temperature.

With installations where cryogens such as liquid nitrogen or liquid oxygen are piped from the liquid receiver to a use point, great care must be given to the design of the piping system. Specifically, where a constant reliable flow is required as is the case for most LN2 freezing equipment.

It should be understood, that as a cryogenic fluid flows through a pipeline intrinsic heat leaks from the external environment will raise its temperature. If the heat gain is large enough and the pipe is of sufficient length, the liquid will change into a two-phase flow. As this mixture is traveling down the length of the pipeline, it absorbs heat from the wall, and its void fraction and average velocity increase as its average density decreases. Generally, by the time the mixture reaches the end of the pipeline, the two-phase wall friction is 30 times larger than that expected in single-phase flow under the same conditions. This phenomenon can be eliminated or greatly minimized by using a piping design that allows practically no heat absorption such as with VJ pipe.

8.3.1   The Geyser Effect

Most often, this very peculiar behavior of LN2 in a pipeline feeding the freezing equipment is not understood and as such no preventive design features are included. The behavior of cryogenic fluids in pipelines containing horizontal and vertical lines with elevated storage tanks is complicated by the geyser effect. This phenomenon, which typically occurs at low flow rates or with stagnant fluids, is characterized by a sudden boiling of all the cryogenic fluid in the vertical pipeline. This process of geysering and refilling will repeat over and over again which can cause water hammer effect and damage to the pipeline. The principle heat gain into these systems comes primarily from the shut off valve at the bottom of the pipe. Experiments (Filina 1983) on the geyser effect have been performed using industrial-scale components.

In cryogenic fluid pipelines, where a main header-pipe supplying several vertical drops to associated equipment is used, the geyser effect can cause unpredictable flow loss to the equipment, specifically the drops at the end of the main header.

How it works

A normal T-drop

With the entire line at saturation temperature, the fluid is entirely in equilibrium and (Qw = constant) now the isolation valve is opened and flow occurs. A rapid down ward flow will now satisfy the demand of the application equipment. Now the isolation valve is closed, either manually or by electro-mechanical means. Almost immediately nucleate boiling begins. In the vertical pipe, a phase called slug flow will occur and will rapidly shoot up and block the flow in the main header at the interface of horizontal and vertical pipe. This in turn will cause dual phasing in the pipeline down stream, starving the other drops.

Using a siphon loop

With the opening of the isolation valve, the pressure in the vertical line will drop creating a siphon effect and subsequent fluid flow. Now, the valve is closed. Again, nucleate boiling will occur. However, due to the loop the vapor created will rise and come to an equilibrium at the line (A) and P1=P2 as depicted in Figure 14.

As long as we have this equilibrium condition, the flow in the main header will not be disturbed.

Fig. 14. The geyser effect: (a) nucleate boiling, (b) Slug flow. (c) Water hammer phenomena. Qo = Heat absorption from valve.

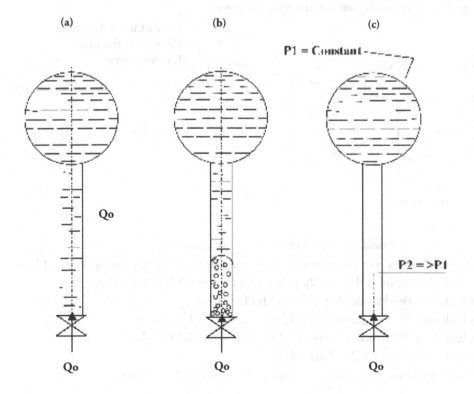

(a)   (b)   (c)

Fig. 15   How to circumvent the geyser effect.

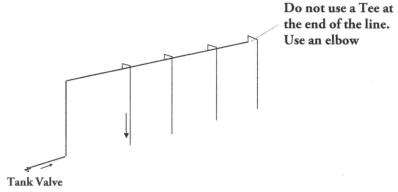

P1 = P2

A — — — — — — — — — — — A

Main Header

Supply Line to Process

WRONG METHOD   **CORRECT METHOD**

Fig. 16   Typical layout for multiple use points

**Do not use a Tee at
the end of the line.
Use an elbow**

Tank Valve

8.3.2   *Economic Evaluation of VJ vs. PU Foam Insulated Pipe*

A case study for a 1" copper line with 4" of insulation and an OD of 10" vs. a VJ line:

Heat of vaporization of LN2 at 20 psig tank pressure = 80.50 BTU/Lb.

Heat leak into rigid static VJ pipe= 0.45 BTU/ft-hr.

Heat leak into PU insulated pipe (NEW) = 20.00 BTU/ft-hr.

Heat leak into PU insulated pipe (OLD) 7 yrs. = 200 BTU/ft-hr.

Assumed cost of the LN2 is $0.04 /Lb.

Hours of operation, 8 Hrs/day x 5 d/wk x 50 wks/yr = 2000 Hrs/yr.

NOTE: The length of pipe in feet is always calculated in equivalent feet. Valves, elbows, etc. must be converted into total linear pipe length. (Use Table 7)

Heat leak of the VJ pipe = (0.45/80.5) x ($0.04) x (2000)= $0.45/ft/yr.
Heat leak of new PU pipe = (20/80.5) x ($0.04) x (2000)= $19.88/ft/yr.
Heat leak loss of old PU pipe = (200/80.5) x ($0.04) x (2000)=$198.76/ft/yr.

The savings can now be tabulated as follows:
VJ pipe vs. new PU insulated copper pipe will be, $19.88 - $0.45 = $19.43/ft/yr.
VJ pipe vs. old PU insulated copper pipe will be, $198.76 - $0.45 = $198.31/ft/yr.

Evaporation of the LN2 in a static line:
PU insulated: n an elevated 1" line (NEW), in about 30 seconds the LN2 vaporizes and is pushed back towards the storage tank.
VJ pipe: The line is beginning to empty after 5 minutes.

8.3.3   When to Select VJ or PU Insulated Pipe

In cryogenic piping systems, two-phase flow is difficult to avoid. We know this and as such when designing these piping systems we must use the best quality possible. Make sure to calculate the pipe size allowing a minimum pressure drop. It takes only a slight localized heat influx and/or pressure drop to cause the evaporation of the fluid. This vapor can significantly reduce the liquid flow and often becomes a factor of a reduced freezing capacity of the freezer.

Do not make a design as shown in Fig.37 it maximizes the pressure drop. The use of an electrical Solenoid Valve (SOV) in a LN2 line is *not* recommended. Fig. 36 is a manifold that produces a minimum pressure drop and should be able to introduce a liquid into the tunnel spray header with very little vapor. The use of a modulating flow control to control the LN2 flow into the spray header is another means of reducing the formation of vapor. This type of control allows the liquid to flow with a relatively constant rate and thus allowing the cryogen to become equilibrated and laminar. Insulate all necessary components with at least 4" of foam encapsulated with a ¼" thick PVC jacket. Fig.36 is a good example of a well-designed manifold.

Here again, pay attention to this detail of the installation, it will decrease the efficiency of the freezer if not done correctly and will directly increase the freezing cost.

Thus, it should be simple to decide which type of piping to use when designing a freezing system. VJ insulated piping is able to deliver a liquid which conforms as closely as possible to the theoretical supercritical state of the cryogen with the maximum BTUs. Since a freezer's efficiency and *freezing cost* are affected by the caloric value of the injected LN2, a VJ insulated piping system is the preferred choice.

If cost becomes an issue, PU insulated LN2 piping systems can be used for short runs, less than 100 ft. equivalent length, a vertical line of not more than 8 feet, and with production rates that are constant and run for at least 8 hours per day. Once the operation becomes routine and a view years have past, replace the piping with a good VJ piping system. Empirical data have shown that an immediate reduction of the freezing cost will be realized.

Fig. 17   CRYOGENIC PIPE DESIGN: HORIZONTAL TO VERTICAL DROPS.

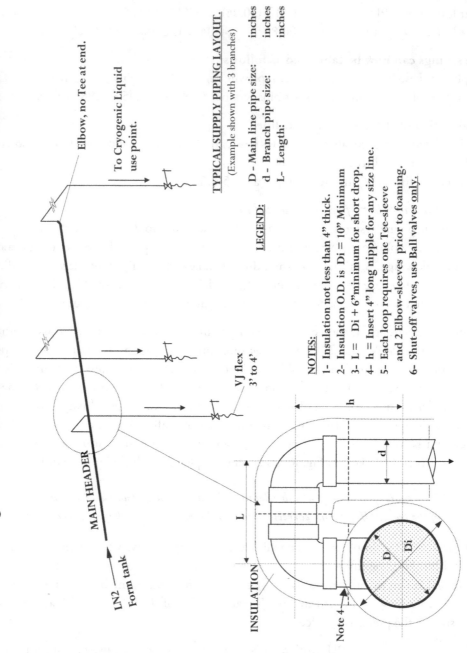

Elbow, no Tee at end.

To Cryogenic Liquid
use point.

VJ flex
3' to 4'

MAIN HEADER

LN2 —
Form tank

INSULATION

Note 4

TYPICAL SUPPLY PIPING LAYOUT.
(Example shown with 3 branches)

LEGEND:

D - Main line pipe size: inches
d - Branch pipe size: inches
L- Length: inches

NOTES:

1- Insulation not less than 4" thick.
2- Insulation O.D. is Di = 10" Minimum
3- L =   Di + 6"minimum for short drop.
4- h = Insert 4" long nipple for any size line.
5- Each loop requires one Tee-sleeve
   and 2 Elbow-sleeves prior to foaming.
6- Shut-off valves, use Ball valves only.

POLYURETHANE & VACUUM JACKETED CRYOGENIC CONDUITS

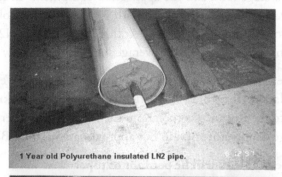

1 Year old Polyurethane insulated LN2 pipe.

Fig. 18

This pipe lost most of its insulating characteristics

5 Year old Polyurethane insulated LN2 pipe.

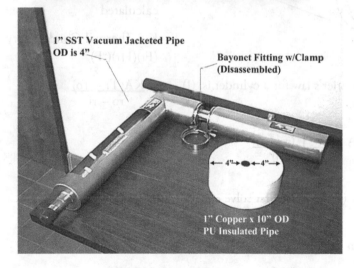

1" SST Vacuum Jacketed Pipe OD is 4"

Bayonet Fitting w/Clamp (Disassembled)

Fig. 19

1" Copper x 10" OD PU Insulated Pipe

8.4.0   Heat Leak into a (LN2) Transfer Line

HEAT: Energy in transition between a "system" and its surroundings that is caused to flow by a temperature difference.

HEAT FLOW: The direction of heat flow is *always* from warm to cold.

The most often encountered type of heat influx into a cryogenic system is the heat leak from the piping between the storage tank and the application equipment. In the following example we will use a ¾" copper type "K" pipe insulated with 3" and 4" thick Polyurethane foam. The fluid is LN2 coming from a 20 psig storage tank. At this pressure the LN2 has a temperature of -306°F. The ambient temperature is 90°F. The average temperature is now -216°F.

We find that a ¾" copper "K" pipe has an outside diameter of 0.875". We will use this value as the (ri) of the *foam* insulation. The (ro) is simply the 0.875" plus the thickness of the insulation, which is 3". According to the Rovanco corporation, the (K) value or Thermal Conductivity for "INSUL-8" at an average temperature of -216°F is approximately 0.0108 BTU/ (Ft) (Hr) (°F).

Now calculate what the heat leak will be per foot of pipe:

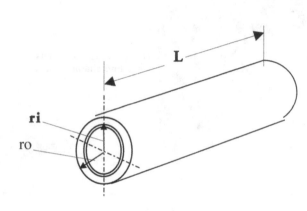

ri = Inside radius in feet = 0.875" = 0.073'

ro = Outside radius in feet = 3.875" = 0.323'

L = Length in feet = 1 foot

Ti = Inside temp. °F = - 216°F

To = Ambient temp. °F = 90°F

A = Insulation area in feet^2 = To be calculated

K = Thermal Conductivity = 0.0108 BTU/ (Ft)(Hr)(°F)

The formula of Fourier's law for a cylinder is: (1) $q = \dfrac{KA(Ti - To)}{ro - ri}$

Where: (2) $A = \dfrac{2\pi \cdot L\,(ro - ri)}{\ln\,(ro/ri)}$

To solve equation (1) we must first solve equation (2) to calculate (A), the area of insulation in square feet.

Solving Equation 2:

$A = \dfrac{2\pi \times L\,(ro - ri)}{\ln(ro/ri)} = \dfrac{(2)(3.14)(1')(0.323 - 0.073)}{\ln(0.323\,/\,0.073)} = 1.056\ \text{FT}^2.$

Solving Equation 1:

$q = \dfrac{KA(Ti - To)}{ro - ri} = \dfrac{(0.0108)(1.056)(-306° - 90°)}{(0.323 - 0.073)} =$ **9.84 Btu/Hr./Ft. (For 3" Thick PU)**

Now let us take the same pipe, however it is insulated with 4" of PU foam insulation used primarily with LN2 piping:

Hence, ri = 0.875" = 0.073' (Inside radius of the insulation)

ro = 4.875" = 0.41' (Outside radius of the insulation)

Ti = -306°F

To = 90°F

K = 0.0108 BTU/(Ft)(Hr)(°F)

Solving Equation 2:

$$A = \frac{(2)(3.14)(1')(0.41 - 0.073)}{\ln (0.41/0.073)} = \frac{2.116}{1.73} = 1.223 \text{ FT}^2.$$

Solving Equation 1:

$$q = \frac{(0.0108)(1.223)(-126°)}{(0.38 - 0.04375)} = -\frac{1.664}{0.34} = \textbf{4.9 Btu/Hr./Ft. (4" Thick PU)}$$

8.4.1   *Calculating the LN2 Pipe Size*

The transfer piping is the second most important component of the freezing system. It will directly influence the freezing cost.

Let's face it; LN2 transfer through piping is complex. From a designer's point of view we can now accommodate this and design the best system we can with the data we have. Remember, our goal is to transfer the LN2 from the storage tank to the freezer with as little vapor as possible, keep it supercritical. The two main concerns for the designer are:

1.   The heat leak into the transfer piping

2.   The total pressure drop

First let's collect the data to calculate the pipe size.

1.   Heat leak: for 4" thick 1" P.U. foam insulated pipe - 21 BTU/h/ft. for 1" VJ pipe - 0.47 BTU/h/ft.

2.   The pressure drop should not be more than 2.9 psig., creates +/- 1% vapor.

3.   At a tank pressure of 20 psig, the heat of vaporization (H_{vap}) of LN2 is 80.5 BTU/Lb.

NOTE: The heat leak data are average numbers. Contact the pipe manufacturer for the relevant numbers.

•   We will use an arbitrary LN2 flow rate of 1400 Lbs/hr. (3.5 GPM)

•   The equivalent transfer line is 90 ft. of copper "K" P.U. insulated pipe.

•   The theoretical % of vapor, dual phase condition, allowed to form is 1%

8.4.2 *Flow Reduction Due to a Dual Phase Condition*

Percentage (mass) of vapor   Reduction ratio = $\dfrac{\text{Real dual phase flow}}{\text{Super critical flow}}$

Percentage (mass) of vapor	Reduction ratio
0 %	1
0.5	0.90
1	0.80
2	0.68
3	0.61
4	0.56
5	0.52
10	0.39

The required LN2 flow can be calculated by dividing it by the reduction ratio number.

The required LN2 flow divided by the reduction factor gives the actual volumetric flow, 1,400 / 0.80 = 1,750 Lbs/hr. or 259.175GPH = 4.32 GPM

The LN2 flow requirement is 4.32 GPM. Select 5 GPM on Table 8

1.   Loss in ft. for a ½" pipe are the average between 4 and 6 GPM or (27 + 57)/2= 42 head ft.
       The pressure required to overcome this head is now (42 x 0.81)/0.433 = 78.6 psig. Obviously, a ½" pipe is too small.

2.   Loss in ft. for a ¾" pipe is (7 + 14.7)/2= 10.85 head ft.
       The pressure required to overcome this head is now (10.85 x 0.81)/0.433 = 20.3 psig.

3.   Loss in ft. for a 1" pipe is (2.14 + 4.55)/2= 3.35 head ft.

       The pressure to overcome this head is now (3.35 x 0.81)/0.433 = 6.3 psig.
       Which pipe size do we select?

◊ If we need LN2 in a Spray Counter Flow or immersion freezer, select a 1" pipe.

◊ A Convection type freezer can use a ¾" pipe.

RESISTANCE OF VALVES & FITTINGS TO FLOW OF -LN$_2$ IN EQUIVALENT LENGTH OF PIPE IN FEET.

(Table 7)

	Nominal Pipe Size						
	3/8"	½"	¾"	1"	1 ½"	2"	2 ½"
Globe Valve	10	15	20	25	42	55	65
Ball Valve	1.0	1.5	2.2	3.0	4.0	5.0	6.0
Solenoid Valve (110 V/N.C.)	13	20	26	33	55	72	85
Tee (Through Side outlet)	2.1	3.3	4.5	5.5	9.0	11.0	14.0
Tee (Straight through)	0.7	1.0	1.5	1.8	2.8	3.5	4.2
Elbow 90°	1.2	1.5	2.0	2.5	4.0	5.1	6.0
Elbow 45°	0.5	0.8	1.0	1.2	2.0	2.5	3.0

Example: Calculating the equivalent pipe length.

A LN2 tank is connected to the freezer by a piping system consisting of 50 ft. of 1" P.U. insulated pipe, (1x) globe valve, (1x) tee, (4x) elbows, (1x) ball cryogenic long stem control valve, (1x) "Y" strainer and a 15' vertical run.

The equivalent pipe length is:

50' + 25' + 1.8' + (4 x 2.5) + 3' + 25' + (15' x .43) = 121.25 ft. (122 ft.)

NOTE: For a "Y" strainer with an *80 mesh* screen, use the values for a Globe Valve.

Friction loss of water in feet per 100 foot length of pipe
Base on Williams & Hazen formula using a constant of 100.
Pipe sizes of standard pipe in inches

Table 8

U.S. GPM	1/2" Vel. (ft/sec)	1/2" Loss (in ft)	3/4" Vel. (ft/sec)	3/4" Loss (in ft)	1" Vel. (ft/sec)	1" Loss (in ft)	1.5" Vel. (ft/sec)	1.5" Loss (in ft)	2" Vel. (ft/sec)	2" Loss (in ft)	2.5" Vel. (ft/sec)	2.5" Loss (in ft)	3" Vel. (ft/sec)	3" Loss (in ft)	4" Vel. (ft/sec)	4" Loss (in ft)
2	2.1	7.4	1.2	1.90												
4	4.21	27	2.41	7.00	1.49	2.14	0.63	0.26								
6	6.31	57	3.61	14.70	2.23	4.55	0.94	0.56	0.61	0.20						
8	8.42	98	4.81	25.00	2.98	7.8	1.26	0.95	0.82	0.33	0.52	0.11				
10	10.52	147	6.02	38.00	3.72	11.7	1.57	1.43	1.02	0.50	0.65	0.17	0.45	0.07		
12			7.22	53.00	4.46	16.4	1.89	2.01	1.23	0.79	0.78	0.23	0.54	0.10		
15			9.02	80.00	5.60	25	2.36	3	1.53	1.08	0.98	0.36	0.68	0.15		
18			10.84	108.02	6.69	35	2.83	4.24	1.84	1.49	1.18	0.50	0.82	0.21		
20			12.03	136.00	7.44	42	3.15	5.2	2.04	1.82	1.31	0.61	0.91	0.25	0.51	0.06
25					9.30	64	3.8	7.3	2.55	2.73	1.63	0.92	1.13	0.38	0.64	0.09
30					11.15	89	4.72	11	3.06	3.84	1.96	1.29	1.36	0.54	0.77	0.13
35					13.02	119	5.51	14.7	3.57	5.10	2.29	1.72	1.59	0.71	0.89	0.17
40					14.88	152	6.3	18.8	4.08	6.60	2.61	2.20	1.82	0.91	1.02	0.22
45							7.08	23.2	4.6	8.20	2.94	2.80	2.04	1.15	1.15	0.28
50							7.87	28.4	5.11	9.90	3.27	3.32	2.27	1.38	1.28	0.34
55							8.66	34	5.62	11.80	3.59	4.01	2.45	1.58	1.41	0.41
60							9.44	39.6	6.13	13.90	3.92	4.65	2.72	1.92	1.53	0.47
65							10.23	45.9	6.64	16.10	4.24	5.40	2.89	2.16	1.66	0.53
70							11.02	53	7.15	18.40	4.58	6.20	3.18	2.57	1.79	0.63
75							11.8	60	7.66	20.90	4.91	7.10	3.33	3.00	1.91	0.73
80							12.59	68	8.17	23.70	5.23	7.90	3.63	3.28	2.04	0.81
85							13.38	75	8.68	26.50	5.56	8.10	3.78	3.54	2.17	0.91
90							14.71	84	9.19	29.40	5.86	9.80	4.09	4.08	2.30	1.00
95							14.95	93	9.7	32.60	6.21	10.80	4.22	4.33	2.42	1.12
100							15.74	102	10.21	35.80	6.54	12.00	4.54	4.96	2.55	1.22
110							17.31	122	11.23	42.90	7.18	14.50	5.00	6.00	2.81	1.46
120							18.89	143	12.25	50.00	7.84	16.80	5.45	7.00	3.06	1.71

Pressure head, in feet, equals 0.433 times the pressure, in psig, divided by the specific gavity of the fluid

Head Pr. $= \dfrac{0.433 \times psig}{SG}$ $psig =$ Head Pr. $\times$ SG

0.43

SG LN2 $= 0.81$

FIG. 20 11000 & 600 GALLON LN2 TANKS.

Figure 20, depicts a typical nitrogen supply system at a food processing plant. The high pressure, 100 psig 600 gallon tank provides nitrogen gas to an automatic packaging line where GN2 is used to package a product to extend its shelve live. Here, the GN2 is used for its inerting propertics.

The 11000 gallon tank supplies a low pressure, 20 psig, LN2 to a "Convection" type freezing tunnel. Notice that the supply line is a P.U. insulated line and not a vacuum jacketed line. The tunnel is located directly on the other side of the wall and operates 10 hrs./day.

Figure 21   30,000 GALLON LN2 RECEIVER

NOMENCLATURE:

1. Safety tree
2. Main Vapor blow-down line – 3"
3. Vapor line from pressure building vaporizer – 2"
4. Main fill line – 1.5"
5. Liquid line to pressure building vaporizer – 0.5"
6. Tank liquid level indicator
7. Telemetry indicator
8. Main fill connection with cap.

9. Pressure controller with,
10. Pressure control vaporizer

The 3" main feed line to the tunnel is located on the opposite side of the tank.

Figure 22   LN2 TANK PRESSURE CONTROL

Depicted is a typical "Constant Pressure Control" device. Static pressure in a LN2 tank varies with the liquid level of the nitrogen. To be able to supply the freezer with a *consistent* quantity of LN2, the pressure in the tank must be kept constant at all levels of LN2. At low tank pressures, 20 psig or less required for cryogenic freezing, the conventional mechanical pressure regulators supplied with the tank are not suitable for this fine control.

8.5   TYPE OF FREEZERS

8.5.1   The "Spray Counterflow" Tunnels
(Figure 24, 24A)

The **Spray Counter flow** freezing tunnel is a sophisticated piece of equipment designed to take full advantage of the energy of Liquid Nitrogen (LN2) as a freezing media and still keep it affordable, simple to operate and maintain. With the storage tank and the piping system designed properly, the LN2 will enter the tunnel super cooled. To be able to monitor this, a pressure gauge up-stream of the modulating valve is provided to indicate the tank pressure (Head pressure plus liquid column) and a pressure gauge down stream of the modulating valve to indicate the pressure drop of the entire piping system. This second gauge should indicate between 2 and 5 PSIG with a fully opened valve. (See Fig. 36 & 37) These modern freezers use LN2 very efficiently with an energy conversion of 85% or better depending on the tank pressure and manufacturer. As illustrated in Table 2, a 100% conversion gives us 159 BTU/Lb. @ 20 PSIG.

8.5.2   The Process

The **Pre-cool Zone** is designed to utilize the Sensible Heat of the LN2 to pre-cool the product as it enters the freezer and travels towards the cryogenic spray zone. It is important to understand why this type of tunnel uses more than 50% of its length for the pre-cool zone. For LN2 to release 100% of its energy, all of the sensible heat of the vapors must be used for the freezing process. Table 2 shows that 50% of the injected LN2 comes from the sensible heat.

$$(H_{vap} + S.H = Q_{Total})$$

Circulating fans impinge the cold vapors over the product to improve the heat transfer rate. It is extremely important that all of the circulating fans are functional to maintain an evenly frozen product and to produce a product with the required end temperature (T2) within the predetermined time. With just one fan out of service, the product will exit the tunnel at a higher temperature as originally calculated. The response is then to lower the freezing temperature or set point temperature of the tunnel, which will disrupt the pressure balance inside the tunnel subsequently increasing the LN2 consumption.

Tunnels longer than 20 feet are equipped with a "Scroll Fan" at the end of the pre-cool zone to assist the flow of the vapors coming of the cryogenic zone to flow towards the circulating fans in the pre-cool zone for a more efficient distribution of the cold vapors. The scroll fan(s) are equipped with a frequency controller to allow fine-tuning of the vapor flow. This operation can be integrated in the control system or it can be manually controlled.

At the entrance of the tunnel an exhaust plenum is installed to direct the flow of the expanded vapors towards the exterior of the building. See Figures 38 to 40. As described in Chapter 10, the exhaust system and its management is what differentiate between an efficient and inefficient tunnel freezer.

The **Spray or Cryogenic Zone** follows the pre-cool zone. Nozzles mounted in a spray header distribute the LN2 directly onto the product. At this point, the temperature and state of the LN2 is dependant on the pressure of the LN2 as it leaves the storage tank. (See Table 2.) Practically it can be as low as -315.6°F at a pressure of 5 psig when a horizontal tank is used. This cold LN2 will vaporize when it contacts the product and the ensuing rapid temperature reduction freezes the product. The interior heat of the product is now conducted towards the colder surface. Some of the LN2 that flows trough the belt is collected in a tray, which is installed under the spray header area. This tray prevents the LN2 from collecting on the bottom of the freezer. This should be avoided at all cost to prevent serious damage to the insulated bottom, which will buckle and crack exposing the interior of the insulated bottom to water and bacteria.

Air movement in the cryogenic zone should be avoided to allow the LN2 to contact the product as a Liquid to get the maximum freezing effect.

The **Equilibration Zone** is the last section of the tunnel and is often called the soak-zone. As implied, it allows the now very cold product surface to come into equilibration with the entire mass of the product. When the vapor balance of the tunnel at full capacity has been properly adjusted, only a very small amount of nitrogen vapor should exit this part of the tunnel. If absolutely no vapor exits this part of the tunnel, too much outside warm air is drawn into the tunnel. This heat requires additional LN2 and in addition will result in excess ice build up in the tunnel, which could damage the conveyor mechanism and circulating fans. *Adjust the scroll fan and exhaust pull to correct this.* A well-balanced tunnel has a small amount of vapor exiting the tunnel. Make this adjustment only when the tunnel has been in full production for about 5 minutes at the correct set point temperature and let it run with the new adjustments for at least 5 minutes before making additional corrections if necessary.

Advantages: *Most efficient tunnel.* Is very user friendly and has the ability to use a range of set point temperatures. This is often useful if a spike in the production requirement is necessary. This type of tunnel can increase its capacity by as much as 50% by reducing its set point temperature for an emergency production run. However, it follows that by doing this, the LN2 consumption will increase.

◊ It is not recommended to run the tunnel at set point temperatures below the temperatures recommended by the manufacturer for extended periods.

Disadvantage: *It is more dependent on a good LN2 supply i.e. low pressure.* Errors in the storage tank design and condition will most likely result in a higher supply pressure at the injection point. This will increase the consumption rate and makes it more difficult to balance the vapor flow in the tunnel.

◊ The correct belt loading is essential with this type of tunnel. A poor belt loading will affect the consumption rate.

◊ Spray Counter flow tunnels shorter than 25 feet are less efficient and should be avoided if the freezing cost is of concern.

Ideal to freeze: All products susceptible to dehydration loss.

Raw products such as meat patties, poultry parts, and all seafood species, meat balls, pine apple segments, corn on the cob.

Cooked or Cooked or par-cooked products placed in *open* plastic trays, such as TV dinners are efficiently frozen in a spray tunnel. When a mechanical spiral freezer has reached its maximum production capacity, the placement of a spray tunnel direct at the *end of the spiral* can boost the lines capacity.

8.5.3   *The Convection Tunnel*
(Figure 26)

Convection: Heat or cold flows in a stream of air that is hotter or colder than what it flows over, around, or through.

The **Convection** or isothermal tunnel utilizes cold "air" velocity to remove heat from the product. The LN2 is injected through a header spraying the cryogen into the air stream of the circulating fans. The cold air stream is now blown over the product freezing it to a desired end temperature. The set point temperature of this tunnel is actually the temperature in the entire freezer. It has no cryogenic zone with temperatures of the LN2 as seen with the Spray Counter flow tunnels. The LN2 is introduced somewhere in the middle of the tunnel. The expanded vapors are exhausted on both ends of the tunnel and require two exhaust systems. Using one blower with dampers in the two plenums is not recommended. Dampers will accumulate ice and snow build-up and will disrupt the flow of the vapors causing a serious safety hazard.

The circulating fans are the key to this tunnels efficiency; make sure all units are running at full capacity if possible. Often, when the product is small, it is necessary to reduce the air velocity in the tunnel to prevent the product from being blown of the belt. This will however, increase the specific consumption rate dramatically. To keep the air velocity in the tunnel as high has possible, product guide rails are installed along both sides of the conveyor belt to keep the product on the belt.

Unlike the spray tunnel, the convection tunnel should not have a 100% belt loading; an 85% belt loading is preferred. This allows the free circulation of the cold vapors through the belt to take place.

Advantages: It usually cost less. It is more efficient for the cooling of products. (End temperatures above 32° F) If precise belt loading configurations are not always possible, the convection tunnel is the choice of freezer.

Disadvantages: It is a "Gas Guzzler", with an energy conversion of at best 70%. Never use set point temperatures lower than recommended by the manufacturer. It can result in structural damage of the freezer. The requirement of two exhaust plenums and blowers makes this type of tunnel difficult to vapor balance. It requires constant attention from the operator.

Ideal to freeze: Any type of cooked product. Bakery items, with or without yeast, and par baked rolls. Whole raw fish and vacuum wrapped products. Crust freezing of ice cream novelties placed directly on the belt prior to packaging, fried products such as French fried potatoes. Pre-wrapped or open pizzas. Sauces and soups packaged in pouches. And cooling of any type of product.

8.5.4   *The Tri-Deck or Triple Pass Tunnels*
(Figure 27)

The **Tri-Deck tunnel** can be designed as a spray tunnel or as a convection tunnel. Which design is best depends largely on the type of product to be frozen and is determined during the initial design phase.

The tri-deck tunnel is fitted with 3 sets of belts stacked one above the other. All three belts are independently driven with variable frequency drives. The product enters on one side on the top belt and exits from the 3rd. and bottom belt on the opposite side. If the product to be crust frozen before it drops to the second belt, a spray header is mounted at the front end of the tunnel to quickly crust the product so it is able to drop down to the next belt without sticking or lumping. It is desirable to layer the product on the second belt to begin the temperature equilibration of the product by convection and conduction. The third belt can be loaded with even more product. Depending on the type of product, a layer of about 3" deep is common.

Circulating fans are mounted on the roof of the tunnel and by some manufacturers on the sidewalls as well.

Exhaust plenums are mounted on both ends of the tunnel and require just as much attention as with the convection tunnel.A second spray header can be installed over the second belt for added capacity and is usually isolated from the first header by a manual valve, which can be opened for the added freezing capacity.

Advantages: It has all of the advantages of a tunnel with a larger capacity within a smaller footprint. It is ideal to IQF small products such as diced poultry, cooked or un-cooked. It is more adaptable to run with higher tank pressures without losing too much of the energy of the nitrogen vapors entering the tunnel. 70% to 80% efficiency is nominal for this type of freezer.

Disadvantages: The product must be able to overcome the tumbling motion when it drops from one belt to the other without deforming.

More seriously, if the product is stacked too high on the 2nd. and/or 3rd. belt(s), product spillage is common and results in product loss and ultimately could damage the bottom conveyor belt.

Ideal to freeze: Cooked diced products such as diced ham, diced poultry, and pizza toppings.

8.5.5 The Spiral Freezer
(Figures 28, 29, 30, 31)

Spiral Freezers are a high capacity, compact alternative to the conventional in- line tunnels. The length of a freezing tunnel is in general limited to 60 feet, capable of freezing about 3000 Lbs/hr. at 480,000 BTU/hr. For a higher production output, the spiral freezer was developed.

The spiral freezer is a convection or isothermal freezing system equipped with a continuous tension-less belt guided around a circular drum which can be solid or spoked. It is used for very large production rates such as 4000 Lbs/hr and higher. The entire belt and drum mechanism is enveloped by 6 to 8 inch thick polyurethane SST clad panels which make up the freezer's enclosure. It is able to maintain a freezing environment as low as – 160°. Its compact size has attracted many large food processors of high priced specialty products. The LN2 consumption is about the same as for a tri-deck tunnel and about 10% to 14% higher as a spray tunnel. Basically, a spiral freezer at full capacity has an exhaust gas temperature very close to the set point temperature. A spiral with a set point temperature of -100°F has an exhaust temperature of about -90°F at best. Figure 28, 29, has the LN2 heat balance calculated for these temperatures.

The liquid nitrogen is introduced into the freezer through spray headers mounted in front of a set of circulating fans. The swirling very cold vapors circulate around the drum and trough the layers of belting to freeze the product. The expanded vapors are drawn away by two exhaust plenums mounted around the entrance opening and the exit opening. The management of the exhaust vapors is strictly necessary to maintain a save working environment. It is impossible to influence the exhaust gas temperature one way or the other.

To IQF raw products, a spray zone is installed at the entrance to quickly crust the product before it enters the spiral freezer to eliminate the otherwise unacceptable dehydration losses. This is required for freezing items such as meat patties and seafood items.

> WARNING: Spiral freezers have two high volume exhaust blowers, 3000 to 6000 cfm, running at full capacity when product is being frozen. At these rates, a vacuum will be drawn in the processing room. For the blowers to function properly, a positive pressure is required. A negative pressure will reduce the capacity of the blowers and eventually nitrogen vapors will increase above the save level in the production area and a hazardous situation will be the result. Make-up air must be provided.

8.5.6   Belt Types
There are two styles of conveyor belts used for cryogenic spirals,

*   Standard radius, such as the Ashworth Omni Grid.
*   Close radius Omni Grid also by Ashworth.

NOTE: Other manufactures are able to supply similar belting.

The standard radius belt is similar as the belts use in cryogenic tunnels. It can bend and curve around with an inside diameter equal to 4.4 times the overall width of the belt.

The close radius belt utilizes a design that requires a set of center links that acts as a pivot. Unlike the standard conveyor, which can bend in either direction, the close radius belt will only curve in one direction. This belt can bend around into a circle with an inside diameter of 2.2 times the belt width. This feature allows the designer to place more belt length within a smaller footprint and reduced floor space. The close radius belt trades floor space for ceiling height to obtain the same belt length.

8.5.7   Product Clearance
To allow the product to be frozen and travel freely through the freezer, the distance between the belt levels must have a certain clearance for the product to travel under. The norm by various manufacturers is 3.5". Higher clearances are available upon request. It should be understood, that the total height of the freezer is increased when the product clearance dimension is increased.

8.5.8   Circulating Fans
To maximize the heat transfer coefficient of the spiral freezer, circulating fans are installed in each available corner of the enclosure. These fans rotate at 1750 RPM and are able to produce a velocity around 28 ft/sec. For small or thin products, it is advisable to have frequency controlled variable speed fans to adapt the air flow to a specific product size preventing the product from

becoming air born. One other method is to direct the airflow away from the product by installing louvers in front of each fan assembly. This is the preferred method; air velocity is required to have maximum heat transfer.

8.5.9   Night Injection Feature

A separate low capacity LN2 injection rate is used to keep the freezer cold during off periods. A switch will deactivate the main injection manifold and activates a low flow injection manifold to maintain the freezer at a set temperature over night or during extended breaks. Covers are provided to close off both product openings to keep all the cold within the freezer. If a PLC is provided, the night injection mode is simply selected with the PLC. A once a week cleaning procedure has been approved by the USDA as long as the box temperature is kept around -10°F or lower and a complete cleaning and wash down sanitation is done once a week. However, check with your inspector anyway to verify specific area laws.

8.5.10   Access Doors

Depending on the size of the freezer, two to four full size access doors are provided for easy entrance to the freezer's interior for cleaning and maintenance.

Spiral Advantages: Can be used to freeze or cool a variety of products.

It is relatively adaptable to higher LN2 pressures, 30 to 50 psig. without losing too much efficiency.

It has a very large freezing capacity, generally 3000 to 10,000 Lbs/hr.

Maximum space required is 12 to 14 feet square.

Disadvantages: Requires a rigid preventive maintenance program. Uses excessive amounts of LN2 when very low (-200° to -250°F) set point temperatures are required. Damage to the bottom panels is unavoidable when LN2 accumulates on it.

The operators must be well instructed to run a spiral freezer.

The exhaust blowers are often freezing up which result in excess nitrogen vapors flowing into the processing room. Use a design similar as shown in Fig. 40 to prevent this.

The installation is complex and often requires the removal of structural obstacles.

8.5.11   The Batch or Cabinet Freezer

A **Batch or Cabinet Freezer** is a low cost low capacity convection type cryogenic freezer. It was developed for the start-up processor in need of a freezer capable to quickly freeze or cool a product. It is manufactured from stainless steel 304 and as insulation, high-density polyurethane foam, 4 to 6 inches, is injected between the stainless steel panels. Internal circulating fans contribute to the high freezing rate of these freezers.

Rack and dolly systems are used to place the product in the cabinet before the door is closed and freezing can commence. Set point temperatures of -160°F are possible.

A 6-inch opening is provided in the top of the unit to extract the spent gases. If the distance to an outside wall is within 10 feet, an exhaust duct is sufficient. Distances over 10 feet require the installation of a small exhaust blower. Follow the recommendations of the manufacturer if you are in doubt.

Batch freezers are offered in many configurations for specific applications. Here follows the most common models offered by the major manufacturers.

8.5.12   The Mini-Batch Freezer

The Mini Batch freezer is the smallest unit offered and used in laboratories, schools and as test units in food processing plants. It measures 36" x 34" x 38" high and contains a rack assembly with 5 to 8 bakery trays each measuring 26" x 18" x ½" high. It is equipped with one circulating fan. It is completely portable and is mounted on casters. The control cabinet is mounted on top or on one side of the box and wired for 110 Volts. A temperature controller will control the temperature of the freezer to the set point temperature and maintains it at that temperature until the timer times out. The product is frozen to a desired temperature by setting a specific freezing time at the specific freezer set point temperature.

8.5.13   The Single Door Batch freezer
(Figure 32 and 33)

The Single Door Batch freezer measures 50" x 39" x 76" high and is designed to hold a removable rack with 20 to 24 bakery trays. The trays are spaced about 4 inches apart and are individually removable. This model has three circulating fans mounted in one sidewall for added freezing efficiency. The set point temperature is adjustable to a minimum of -160°F. This model is not portable but is permanently placed on the production floor.

The LN2 consumption depends on how diligent the operator is when he (she) removes one finished rack with product and replaces it with a new one. The longer this process takes, the higher the consumption will be. An average of 110 Btu's can be extracted from one pound of LN2.

8.5.14   The Double Door Batch Freezer

The double door batch freezer measures 120" x 48" x 76" high and is nothing more than two single door units sharing one wall with two independent controls and exhaust openings.

8.5.15   The Twin Compartment Batch Freezer with Pre-cool Cabinet

This model is designed to use the sensible heat of the injected liquid nitrogen to get a better efficiency freezer. The freezing process begins with both compartments loaded with a full rack of product. One compartment begins the freezing process by injecting LN2 into the freezer via a spray header mounted in the circulated air stream. It will inject LN2 until it reaches the desired set point temperature after which it cycles on and off to maintain that temperature. The spent vapors are not exhausted but channeled through the second cabinet compartment pre-cooling the product before it is expelled to the exterior of the processing room. When the first cabinet times out, it signals the operator to empty that compartment and replace it with a new rack of products. After the door is closed, the second compartment begins its LN2 freezing cycle and uses the spend gases to pre-cool the just loaded warm product in compartment number one, etc. The LN2 efficiency of this model is about the same as the spiral freezer at 80% efficiency of QTotal. (See Table 2.) or 126.88 BTU's per pound of LN2.

8.5.16   *The Immersion Freezer*
(Figure 34, 34A)

The LN2 immersion freezer consist of an endless belt conveyor configured to run through the liquid nitrogen which is contained in a well insulated tank. A frequency controller is used to control the belt speed. The LN2 level is controllable within millimeters from the control panel. It is important to be able to maintain a very specific LN2 level above the conveyor belt. This liquid level must be adjusted to correspond with the product thickness or type of product. The correct LN2 depth is selected as a function of the (1) product thickness and (2) the kind of product and its thermal conductivity.

1.   A product with a well defined thickness and fed into the immersion freezer laying flat on the belt is frozen with a LN2 depth not more than the total product thickness. Total immersion is not necessary and will only increase the consumption rate. The vigorous boiling effect of the liquid covers the top of the product sufficiently to completely crust freeze it. ◊ The immersion time is determined by the *length* of the horizontal section of the belt, not the depth of the LN2 above the belt.

2.   Very small particles are frozen differently as just described. Products such as peas or pop corn shrimp are best frozen by immersing them completely in the LN2 at a high belt speed. This will prevent the product from sticking to each other and exiting the freezer in a lumpy mass. A preferred feed method is to drop the product into the LN2 from an opening in the top of the freezer.

The immersion freezer can be used as a stand alone freezer but its high consumption rate has contributed to the development of freezers which can utilize the sensible heat of the nitrogen vapors. The immersion-post-cool freezer for example as depicted in Figure 35.

> ***ATTENTION:** The design of the immersion freezers exhaust, stand alone or with a post-cool freezing attachment, requires some special attention. After all, the exhaust vapors are close to the boiling point of LN2 at -320°F. It is therefore not unusual that the exhaust ducting will freeze shut after a view hours of running. The avoid this; an electrical heating element is installed in the duct just after the immersion's exhaust plenum as shown in Fig. 25. The size of the heater depends on the volume of vapors that will be boiled off the LN2 bath during immersion. When we produce a product at a rate of 1000 Lbs/hr. for instance, and the product requires 80 BTU/lb. to finish a specific freezing task, the total amount of BTUs we need, equals 80,000 BTU/Lb./hr. We know that the Heat of Vaporization of LN2 at atmospheric pressure is equal to 85 BTU/Lb. It follows now that the 80,000 BTUs will vaporize 80,000 / 85 = 941 Lbs of LN2 per hour.*

Now we have to heat the boiled off vapors to about 34°F to avoid forming ice within the duct. The boiled off vapors have a temperature of about -300°F. To heat the 941 Lbs of vapors from -300°F to +34°F, we need {-300 – (+34)}.252 = 84.2 BTU/Lb. or 941 x 84.2 = 79,233 BTU/Hr. = 1321 BTU/min.

1 KW = 57 BTU/min. The rate of required energy is than 1321/57 =23.2 KW/min or 1392 KWH. Select a 24 KW heating coil to prevent the exhaust duct from freezing.

◊ Another excellent method of preventing the exhaust ducting to freeze shut, is the double walled duct with warm air circulation. See Figures 39 & 40.

The warm air circulation intake must be warm air of at least 50°F to make it function satisfactory. In colder climates this is difficult to accomplish during the winter months and some other solution is recommended. If the production area in which the freezer operates is refrigerated, the temperatures are often as low as 45°F. This environment is of course not suitable to be used as a warming circulating air for two reasons. Firstly, it is to cold and has little effect to de-ice the exhaust duct. Secondly, the refrigerated air pulled out of the production area requires additional energy to maintain this low temperature with additional make-up air.

It is therefore practical to use a combination of the two methods for the northern hemisphere.

Fig. 23 VARIOUS LN2 FREEZING SYSTEMS

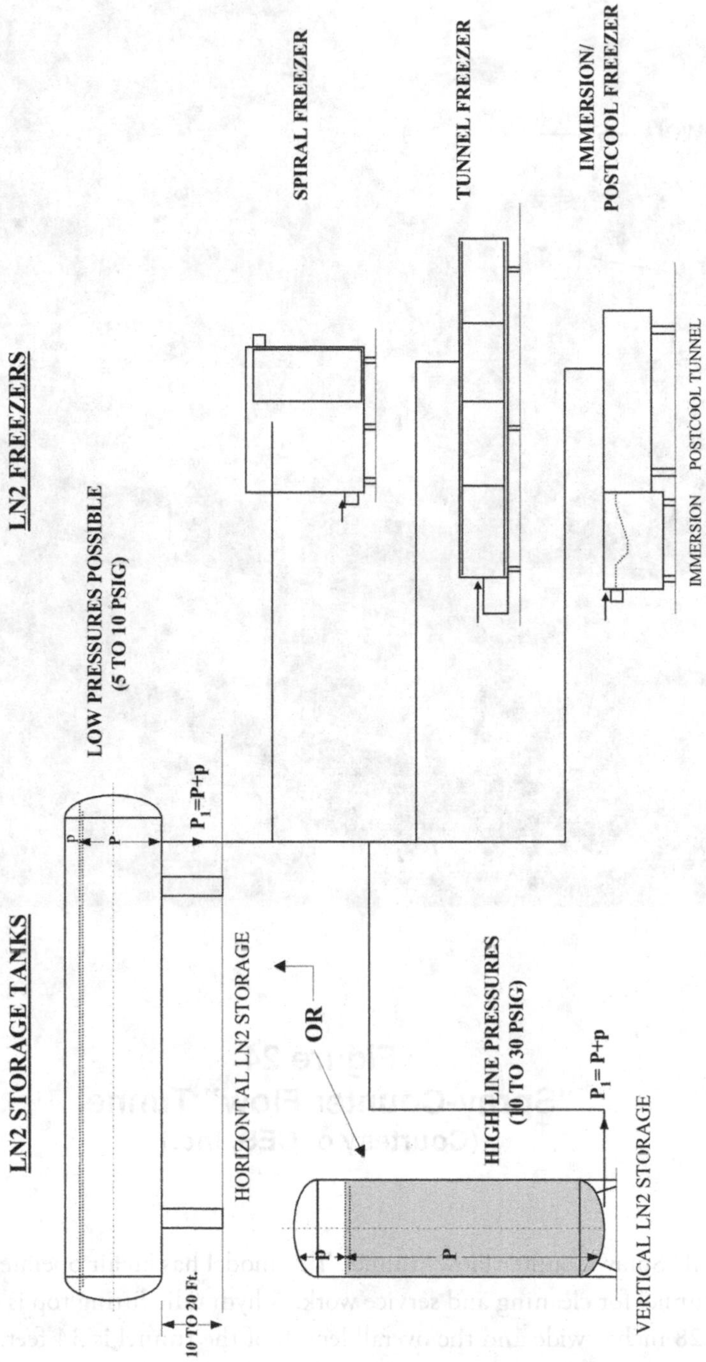

LN2 FREEZERS

LN2 STORAGE TANKS

SPIRAL FREEZER

TUNNEL FREEZER

IMMERSION/
POSTCOOL FREEZER

IMMERSION   POSTCOOL TUNNEL

LOW PRESSURES POSSIBLE
(5 TO 10 PSIG)

$P_1 = P+p$

HORIZONTAL LN2 STORAGE

10 TO 20 Ft.

OR

HIGH LINE PRESSURES
(10 TO 30 PSIG)

$P_1 = P+p$

VERTICAL LN2 STORAGE

Figure 24
"Spray-Counter Flow" Tunnel
(Courtesy of CES, Inc.)

This is a typical "Spray Counter Flow" tunnel. This model has an air operated dropping bottom to access the tunnel for cleaning and service work. A hydraulic lifting top is offered as well. The shown belt is 28 inches wide and the overall length of the tunnel is 34 feet. The load deck is 2 feet long and is followed by the exhaust plenum with a slide that can be adjusted to allow the free passing of the product. One exhaust blower is mounted directly behind the exhaust plenum and is pre-wired into the freezer's control cabinet.

NOTE: This model tunnel is equipped with a top and bottom exhaust.

Fig. 24A   TYPICAL LN2 "SPRAY – COUNTERFLOW" TUNNEL

HEAT BALANCE, LN₂ :
-160°F Set Point Temp. in tunnel.
Tank pressure, 20 psig
Exhaust gas temp. to stack -20°F.
$Q_T = (80.5) - \{ -306° - (-20°) \} \cdot 255 = \underline{153.43}$ BTU/Lb

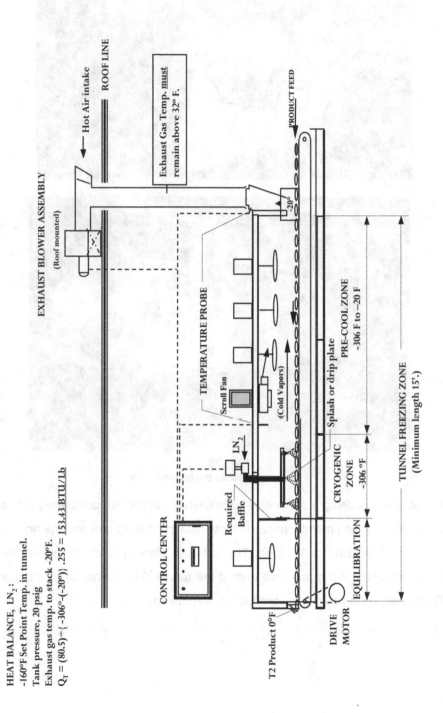

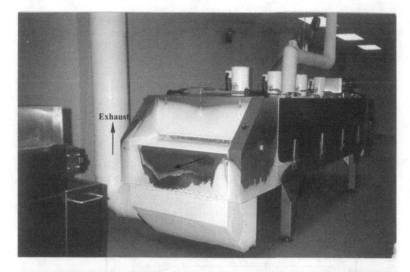

Fig. 25
Courtesy Martin Baron Inc.

This model tunnel is designed with side panels which drop down completely for easy access for cleaning and maintenance. The overall length is 20 feet and the belt is 32" wide. The freezing zone is 15' long. The exhaust plenum is of the below the belt design and installed at the entrance end of the tunnel. The tunnel can be expanded with 10' long sections when more production is required.

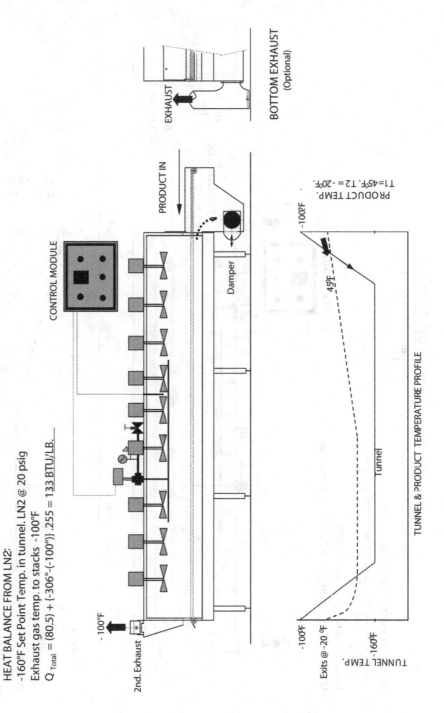

Fig. 26   LN₂ CONVECTION TUNNEL
(Requires 2 Exhausts)

HEAT BALANCE FROM LN2:

-160°F Set Point Temp. in tunnel. LN2 @ 20 psig

Exhaust gas temp. to stacks -100°F

$Q_{Total} = (80.5) + \{-306°-(-100°)\} .255 = 133\ BTU/LB.$

CONTROL MODULE

PRODUCT IN

Damper

2nd. Exhaust

-100°F

EXHAUST

BOTTOM EXHAUST
(Optional)

PRODUCT TEMP.
$T1=45°F, T2=-20°F.$

-100°F

45°F

Tunnel

TUNNEL & PRODUCT TEMPERATURE PROFILE

-100°F

Exits @ -20 °F

-160°F

TUNNEL TEMP.

Fig. 27   LN$_2$ "TRI-DEK" TUNNEL
(Requires 2 Exhausts)

HEAT BALANCE FROM LN2:

-160°F Set point Temp. in tunnel. Tank pressure
@ 20 psig.

Exhaust gas temp. to stacks -50°F.

$Q_{Total} = (80.5) + \{-306° - (-50°)\}.255 = \underline{146\ Btu/lb.}$

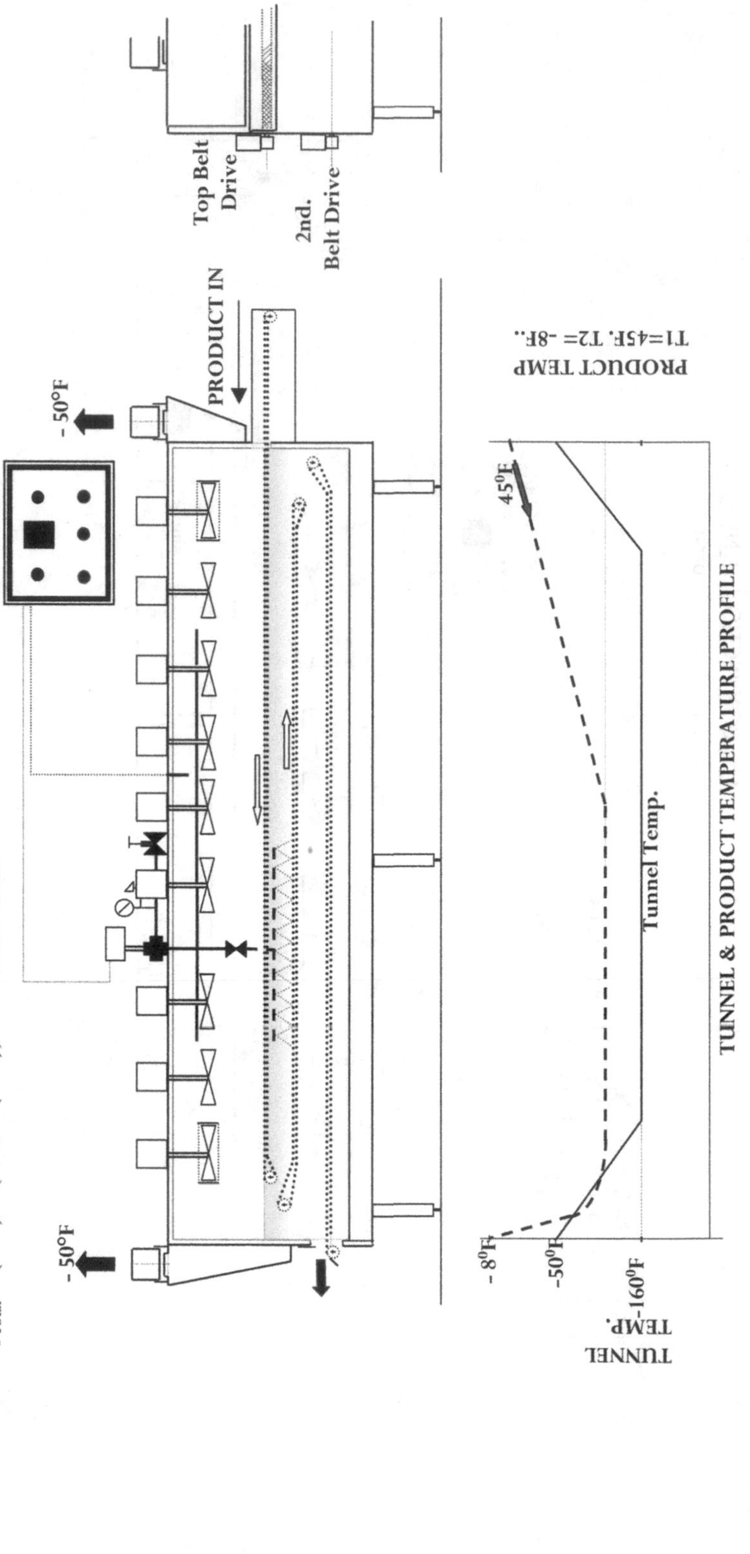

CONTROL MODULE

PRODUCT IN

- 50°F

- 50°F

Top Belt
Drive

2nd.
Belt Drive

PRODUCT TEMP
T1=45F. T2= -8F.

45°F

Tunnel Temp.

- 8°F

- 50°F

-160°F

TUNNEL
TEMP.

TUNNEL & PRODUCT TEMPERATURE PROFILE

Fig. 28   LN2 SPIRAL FREEZER WITH SPRAY BARS

HEAT BALANCE FROM LN2:
-100°F set point temperature
LN2 tank at 20 psig
Exhaust at -90°F
$Q_{total} = (80.5)+\{-305.7°-(-90°)\}.255 = 135.5$ BTU/Lb.

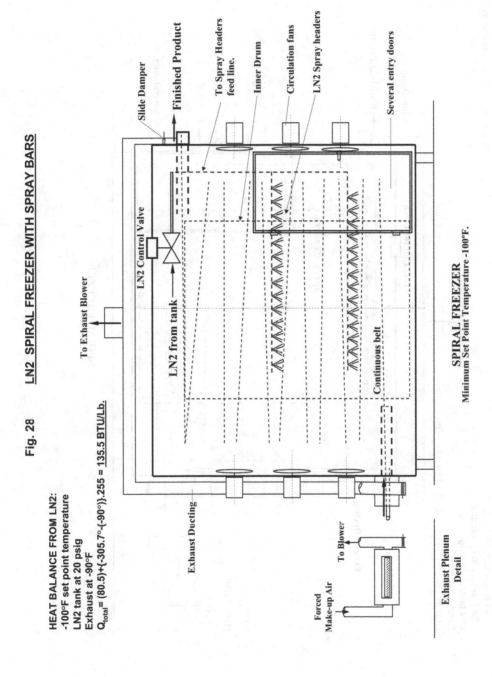

SPIRAL FREEZER
Minimum Set Point Temperature -100°F.

Fig. 29   CONVECTION LN2 SPIRAL FREEZER

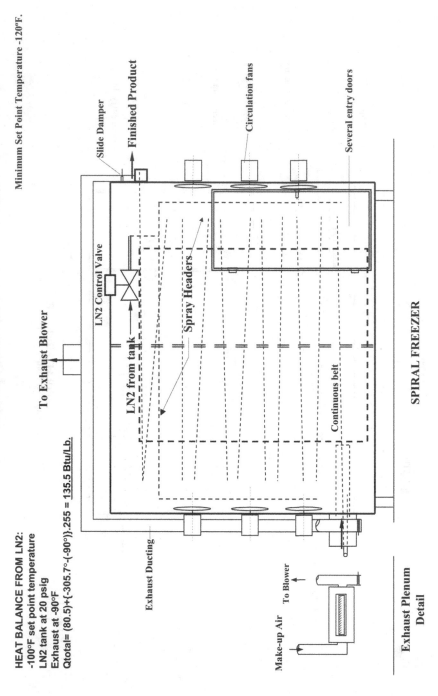

HEAT BALANCE FROM LN2:
-100°F set point temperature
LN2 tank at 20 psig
Exhaust at -90°F
Qtotal= (80.5)+{(-305.7°-(-90°)}.255 = 135.5 Btu/Lb.

Minimum Set Point Temperature -120°F.

To Exhaust Blower

Slide Damper
Finished Product
Circulation fans
Several entry doors
LN2 Control Valve
LN2 from tank
Spray Headers
Continuous belt
Exhaust Ducting

SPIRAL FREEZER

Make-up Air   To Blower

**Exhaust Plenum
Detail**

Figures 30A & 30B   Interior of an 11 Tier Spiral Freezer.

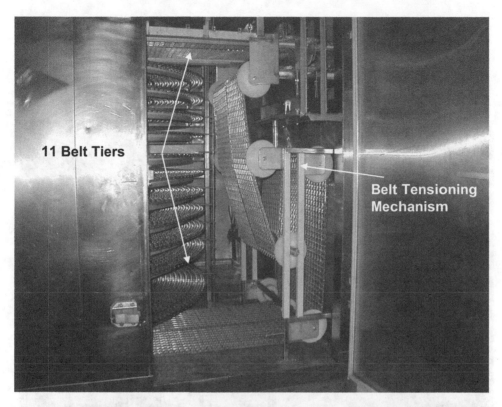

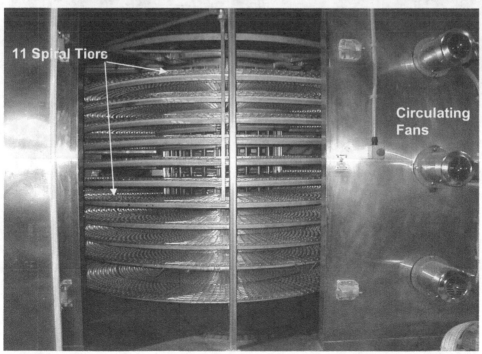

Fig. 31 Two Cryogenic spiral freezers

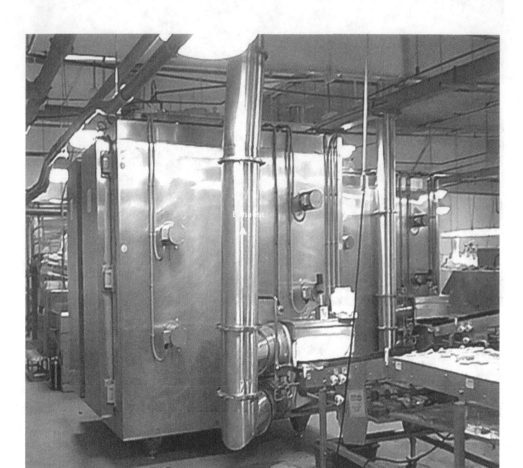

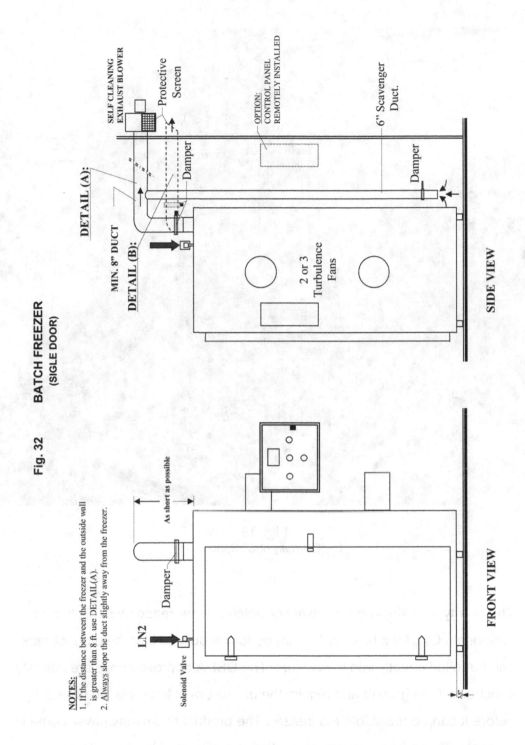

Fig. 32   BATCH FREEZER
(SIGLE DOOR)

NOTES:
1. If the distance between the freezer and the outside wall is greater than 8 ft. use DETAIL(A).
2. Always slope the duct slightly away from the freezer.

SELF CLEANING
EXHAUST BLOWER

Protective
Screen

OPTION:
CONTROL PANEL
REMOTELY INSTALLED

6" Scavenger
Duct.

DETAIL (A):

Damper

MIN. 8" DUCT

DETAIL (B):

Damper

2 or 3
Turbulence
Fans

SIDE VIEW

As short as possible

Damper

LN2

Solenoid Valve

FRONT VIEW

Fig. 33
(Courtesy of CES,Inc.)

This is a typical single door cabinet or batch freezer made by CES, Inc. of Cincinnati, OH. for a bakery. The flat bottom is built so that the product rack can be rolled directly into the cabinet. The USDA approved units are actually 4 inches off the ground and require the product rack to be placed on a dolly before it can be rolled into the freezer. The product racks may never come in contact with the floor. Internal circulation is maintained by 3 circulating fans. BTU's from LN2 @ 20 psig = 110. At 56 psig = 102 BTU's/lb.

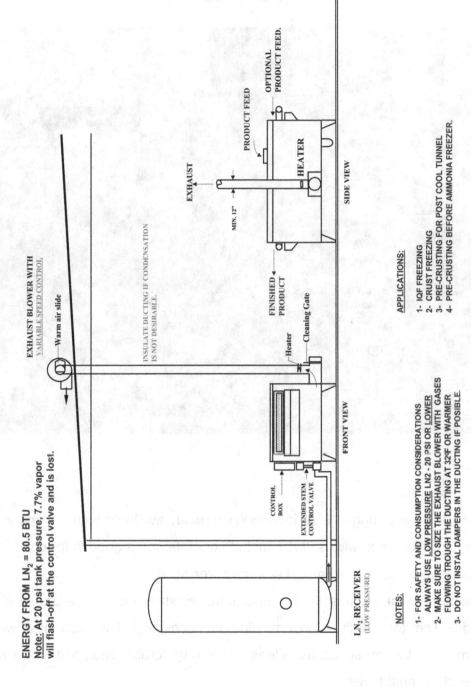

Fig. 34 TYPICAL "IMMERSION FREEZER" INSTALLATION

ENERGY FROM LN₂ = 80.5 BTU

Note: At 20 psi tank pressure, 7.7% vapor will flash-off at the control valve and is lost.

EXHAUST BLOWER WITH
VARIABLE SPEED CONTROL

Warm air slide

INSULATE DUCTING IF CONDENSATION
IS NOT DESIRABLE.

PRODUCT FEED

OPTIONAL
PRODUCT FEED.

EXHAUST

HEATER

MIN. 12"

SIDE VIEW

FINISHED
PRODUCT

Cleaning Gate

Heater

FRONT VIEW

CONTROL
BOX

EXTENDED STEM
CONTROL VALVE

LN₂ RECEIVER
(LOW PRESSURE)

APPLICATIONS:

1- IQF FREEZING
2- CRUST FREEZING
3- PRE-CRUSTING FOR POST COOL TUNNEL
4- PRE-CRUSTING BEFORE AMMONIA FREEZER.

NOTES:

1- FOR SAFETY AND CONSUMPTION CONSIDERATIONS
   ALWAYS USE LOW PRESSURE LN2 - 20 PSI OR LOWER
2- MAKE SURE TO SIZE THE EXHAUST BLOWER WITH GASES
   FLOWING TROUGH THE DUCTING AT 32ºF OR WARMER
3- DO NOT INSTAL DAMPERS IN THE DUCTING IF POSIBLE.

Fig. 34A Immersion Freezer
(Courtesy of MBI, Inc.)

Tee w/cleaning slide

This is a very rugged LN2 immersion freezer, available with a variety of belt configurations and widths. It is simple to operate and a good fit for crust freezing a product before a mechanical ammonia freezer.

Notice how the exhaust duct is erroneously installed without a proper cleaning slide. The tee with a bottom slide is necessary! The duct size with any immersion freezer should be at least 12" and preferably equipped with a variable speed exhaust blower.

CAUTION: Use an exhaust blower designed for cryogenics only!

Fig. 35 **COMPACT IQF TUNNEL – 3,000 LBS/HR. MODEL**
(20'x 3'x 3 tier)

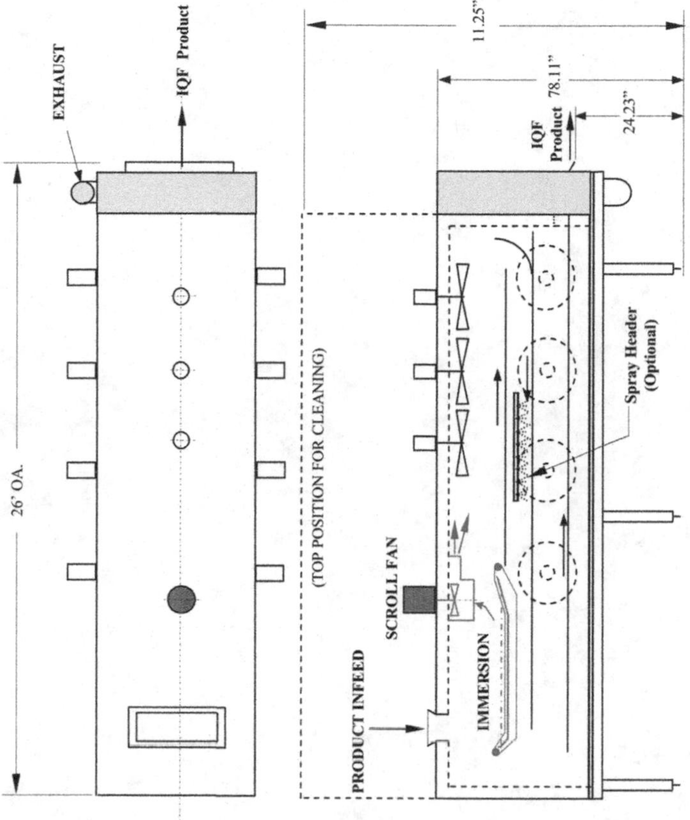

ADVANTAGES:

1. Small foot print with large capacity
2. Efficient LN2 utilization of
   150 to 155 BTU/LB.
3. Front loading feed available

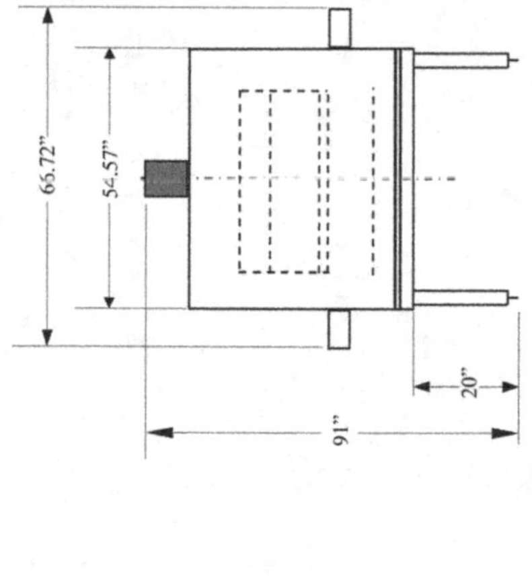

EXHAUST

IQF Product

26' OA.

(TOP POSITION FOR CLEANING)

11.25"

78.11"

IQF Product

24.23"

Spray Header
(Optional)

SCROLL FAN

PRODUCT INFEED

IMMERSION

65.72"

54.57"

91"

20"

9. LN2 CONTROL MANIFOLD

9.1 The temperature controller

General Information

A digital Temperature Controller is mounted in the freezers control panel front cover, directly visible by the operator.

1. Process Value Indicator (˚F)
2. Set Point value Indicator (˚F)
3. Alarm Indicators
4. Mode Keys (Behind cover)

1. Process Value Indicator. (RED)

Displays the process temperature value at that moment and various messages according to a specified display level.

Displays an error code when an error occurs in the temperature controller

Displays "000" for about 4 seconds on power application.

2. Set Point Value Indicator. (GREEN)

Displays the Set Point Temperature and various set values.

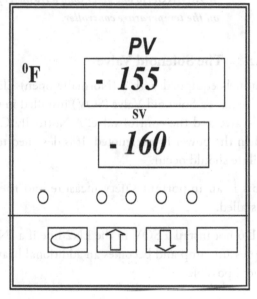

Front View

3. Alarm or Output Indicators.

These indicator lights display various functions such as when an output signal is being transmitted etc. (Study the instruction manual before attempting to use the freezer)

4. Mode Keys

Some manufacturers install these buttons behind a protective panel.

The buttons are used to raise or lower the Set Point Temperature and to change the display modes.

NOTE: A comprehensive manual is furnished with the freezer.

The Temperature Controller is a microprocessor and its primary function is maintaining the "set point" temperature. It does this by constantly comparing the "set point" temperature with the "Real Time" tunnel temperature. The controller receives the tunnel temperature reading from the thermocouple or a RTD sensor located inside the tunnel. *A damaged thermocouple or RTD sensor will render the entire system useless and will shut down the freezing operation.*

In response to the "set point" temperature and the interior tunnel temperature, the temperature controller sends a signal to the LN2 control valve to bring the tunnel temperature to the selected set point temperature.

The freezer is designed to hold a specific set point temperature even though the product load may vary.

> **WARNING: *No one other than a trained operator should be allowed to change settings on the temperature controller.***

9.2   The Solenoid Valve

Tunnels equipped with a (Normally open) Electric Motor driven or pneumatic LN2 control valve have a Solenoid Valve (SOV) installed in the nitrogen supply line between the main shut-off valve and the control valve. A Normally Closed SOV is electrically opened and will close when the power is interrupted. It is designed to close and stop the flow of LN2 when a power failure should occur.

This is an important safety measure and the SOV should be inspected regularly if one is installed.

◊ Do not install a SOV in the LN2 line if a (NC) electric operator is used. The SOV produces a pressure drop and becomes an additional heat source to the LN2 in the pipeline. Avoid using one if possible.

9.3   The LN2 Control Valve

The control valve is actually an extension of the temperature controller. It receives electrical impulses (4 to 20mAmps) from the temperature controller to open or close. There are pneumatic air operated valves and there are electrically operated valves. The pneumatic air operated valves require an air supply to manipulate the valve seat up and down for flow control. This type of valve is often placed in series with a solenoid valve to shut off the LN2 flow in case of a power failure. Avoid this type of installation. The SOV is an additional heat source to the LN2 and it will exacerbate the two-phase condition of the LN2 in the pipeline. Instead, specify a (NC) electrical operator with an extended stem ball valve and characterized seat. This assembly will reduce the heat input to the LN2 in the supply line to the freezer.

9.3.1   Valve Size Does Matter

Another point to consider when selecting a valve is its size. Valve size often is described by the nominal size of the end connection. For most fluid systems, however, a more important measure is the valve's maximum flow rate. The principles of flow calculation dictate that certain design features of the valve are known. These are:

- Size and shape of the orifice and flow path
- Internal diameter of the pipe
- Characteristics of the fluid, such as density and temperature
- Pressure drop from inlet to outlet of the valve

A straight-through flow path, like that of a ball valve, would allow for a greater flow than an equivalent-size needle or globe valve, which presents a much more tortuous flow route. Rather than doing complex calculations to gain a clear understanding of flow, compare the *flow coefficient* (Cv), which incorporates the combined effects of all the flow restrictions in a valve and gives a single common reference number. The following formula is used with the flow coefficient (Cv) to determine the flow rate capability of a valve: **Q = Cv x$\sqrt{}$ ΔP/SG.**

Where, Q is in GPM, P is in absolute pressure, ΔP = pressure differential across the valve, $\sqrt{}$ = square root and SG LN2 = 0.809. Water = 1.

Other valve design features to consider include an on-off or modulating actuation method. Valves with integral end-connections minimize potential leak points and are less labor intensive to install and maintain.

Always use a full port ball valve with an extended stem in LN2 lines.

9.3.2   Installation Procedures

After selecting the right valve for a specific application, consider its proper installation, and look for features that maximize performance and minimize maintenance. Improper installation will affect performance and reliability. Consider these suggestions:

- Valve mounts should handle external loads, such as system expansion, and should absorb torque from valve actuation so that stress is not transferred to the end connections or piping
- Install valves so that they are supported by valve mounting brackets and not by the piping system
- Install valves so that they are easy to see and operate
- Install cryogenic extended stem valves on a vertical plain or according the manufacturers recommendations
- Install valves with the flow in the direction of the flow arrow

9.4   The Thermocouple Sensor

The thermocouple is a thermometer. It is installed through the freezer top and wired through conduit directly to the temperature controller. Do not use aluminum conduits.

A second thermocouple probe farther towards the end of the tunnel is often installed to read the average freezer temperature. (Good feature)

A thermocouple is a temperature measurement sensor that consists of two dissimilar metals joint together at one end (The junction) that produces a small thermoelectric voltage when the junction is heated or cooled. The change in thermoelectric voltage is interpreted as a change in temperature and displayed on the controller. There are many different types of thermocouples based on the type of the dissimilar metals used. Each combination functions best within a specific temperature range. For cold temperature measurements in cryogenic freezers, use a thermocouple indicated by the letter (T) having the dissimilar metals COPPER & CONSTANTAN and sensitive between the temperature ranges of (-350° to +750°F). The wires are insulated with color coded insulation. The copper wire is blue and the constantan wire is red. When using extension wires, the same color coded wires must be used. Both probe wires are mounted in a protective Stainless Steel sheath and are touching the tip of the sheath.

Since thermocouples measure wide temperature ranges and are relatively rugged and inexpensive older cryogenic freezers are equipped with thermocouples. They are used for industrial process applications where accuracy may be a less important factor.

9.5   The Rtd Temperature Sensor

RTD or "Resistance Temperature Detectors" are sensors that measure temperature by correlating the resistance of the RTD element with a temperature. Most RTD elements consist of a length of fine coiled wire (normally platinum) wrapped around a ceramic or glass core, requiring an external power source.

> *RTDs are generally considered among the most accurate temperature sensors available. In addition to offering very good accuracy, they provide excellent stability and repeatability. RTDs also feature high immunity to electrical noise and are, therefore, well suited for applications such as cryogenic freezers.*

9.6   The Set Point Temperature

Selecting the correct "set point" temperature for the tunnel depends on the initial product temperature (T1), the desired equilibrated final product temperature (T2) and the physical nature of the product it self. When the ΔT is large, select a set point temperature close to the minimum recommended temperature of −160°F. With a product having a low T1, a higher set point temperature could suffice.

Be aware however, not all products should be frozen with a "high" set point temperature. With all raw products, freezing should be accomplished rapidly and with cryogenic temperatures. This will produce a good quality product with little or no freezing damage.

For a spray tunnel always keep the set point temperature at a level so that production rates are satisfied but never lower than −160°F. As previously described, for a spray tunnel with a set point temperature of -160°, this temperature represents the average temperature in the tunnel.

> *If −160°F is not getting you the required production rate and all other requirements are met, the tunnel is not long enough!*

Lower than -160°F temperatures are possible for the convection tunnels. However, first consult the equipment manufacturer. Serious structural damage may occur if lower temperatures are used against the manufacturer's recommendations.

◊ When making temperature adjustments, wait until the freezer is running with the maximum production rate for at least 5 minutes. Make each subsequent temperature correction with 5 minute intervals.

◊ The operator should be aware that the exhaust temperature is closely related to the freezer's set point temperature; hence it is necessary to re-adjust the exhaust rate after each temperature change.

> *REMINDER: If maximum moisture retention of the product is desirable during freezing, select the lowest temperature as the set point temperature; such as -160°F and use a spray counter flow tunnel.*

9.7 A "Hot" Vs. A "Cold" Tunnel Freezer

The placement of the temperature probe can be used to design a "Hot" or a "Cold" tunnel. When the probe is installed close to the cryogenic zone of a spray tunnel it will operate with an average higher or "Hot" tunnel temperature. On the other hand, when the probe is installed closer towards the entrance of the tunnel, it will operate with an average lower tunnel temperature or colder tunnel.

Most manufacturers install the temperature probe at a point close to the cryogenic zone and calibrate the average tunnel temperature to be around -160°F hence, a set point temperature of -160°F.

The installation of two probes, one close towards the cryogenic zone and one closer towards the entrance of the tunnel allows the operator to observe at what temperature the tunnel operates at all times. The controller's software will calculate the average internal tunnel temperature and will display that temperature on the controller. In addition, a selector switch gives the operator the option to run the tunnel "Hot" or "Cold" by switching to the individual corresponding temperature probes.

Why is it useful to have these options? For instance, when the tunnel is used to freeze a cooked hot product, it is recommended to design the tunnel as a "Cold" tunnel with the probe closer towards the entrance of the tunnel. It follows that the tunnel is best configured as a "Hot" tunnel when a relatively cold product is processed. In any case, maintain an exhaust gas temperature as warm as possible.

9.8   The LN2 Supply Manifold

The LN2 supply manifold consists of components necessary to deliver the nitrogen to the spray header of tunnel freezers which in turn spray LN2 on the product. This assembly should be designed so that a minimum of heat is transferred to the liquid in the pipe line to avoid excess vapor formation. As illustrated in Fig. 36, three main components are all it takes to properly control the flow into the freezer and maintain an accurate set point temperature.

Fig. 37 is an illustration how not to design a LN2 control manifold.

A CONVECTION spiral freezer has a different method of introducing the LN2 into the freezer. Since a spiral freezer is a convection freezer, it is not necessary to spray LN2 directly on the product. Here, the LN2 is piped to one header in front of each bank of circulating fans. The air velocity of the fans will defuse the nitrogen into the freezer to form a homogeneous temperature within the enclosure.

The SPRAY-CONVECTION combination spiral freezer combines the two methods of freezing by mounting LN2 spray headers over a number of belt tiers.

9.9   Using A PLC To Run The Freezer

A "Process Logic Controller" is actually nothing more than a computer programmed to control mechanical devises. If installed with a cryogenic freezer, all functions of the tunnel can be adjusted on the PLC screen with a key board similar to a lap top. Once programmed, the freezer is running practically on automatic pilot and can be left alone until the end of the production run. It will maintain all parameters as entered in its memory. Dwell time, belt speed, exhaust draw related to the production rate, set point temperature, etc. Using a PLC will greatly improve the freezer's efficiency and result in the lowest LN2 consumption possible.

The use of a PLC is recommended to get the best LN2 consumption rate.

Fig. 36   LIQUID NITROGEN TUNNEL CONTROL VALVE ARRANGEMENT
(OPERATOR: ELECTRIC)

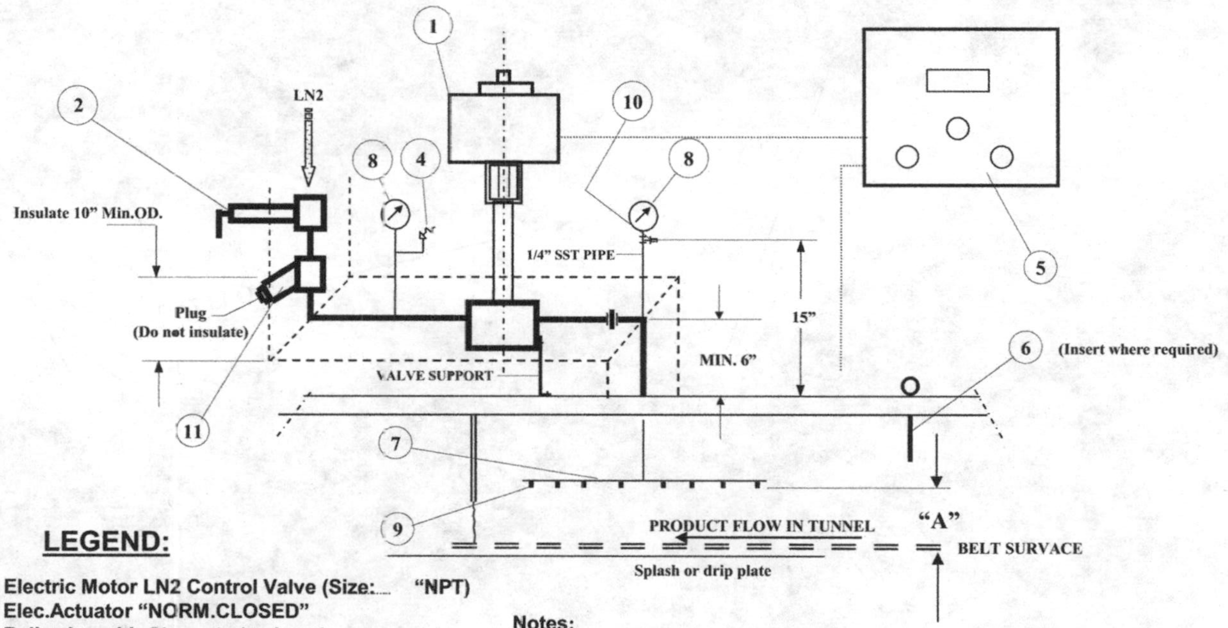

LEGEND:

1- Electric Motor LN2 Control Valve (Size:___ "NPT)
   Elec.Actuator "NORM.CLOSED"
   Ball valve with Characterized seat

*2- Main LN2 shut-off valve (C44 Brass, SST Trim)
   (Ball valve with extended stem)

3- Pr.Gauge - 0-100 PSI

4- Safety, 150 PSI (100 PSI on LN2 Line at the Tank)

5- Temperature Control Module

6- Temperature Probe

7- LN2 Spray Header

8- Tank Press. Gauge - 0-100 PSI (Max.)

9- Spray Nozzles

10- 1/4"SST Tee with plug.

11- Strainer. Brass with SST 80 mesh screen.

Notes:

1. Insulate with 4" thick PU and white 10" PVC Jacket - Install after pressure test.

2. "A" = 6   Inches. (For spray tunnels only)

3. "Characterized" seat – A term used by Worcester Control for modulating ball valves.

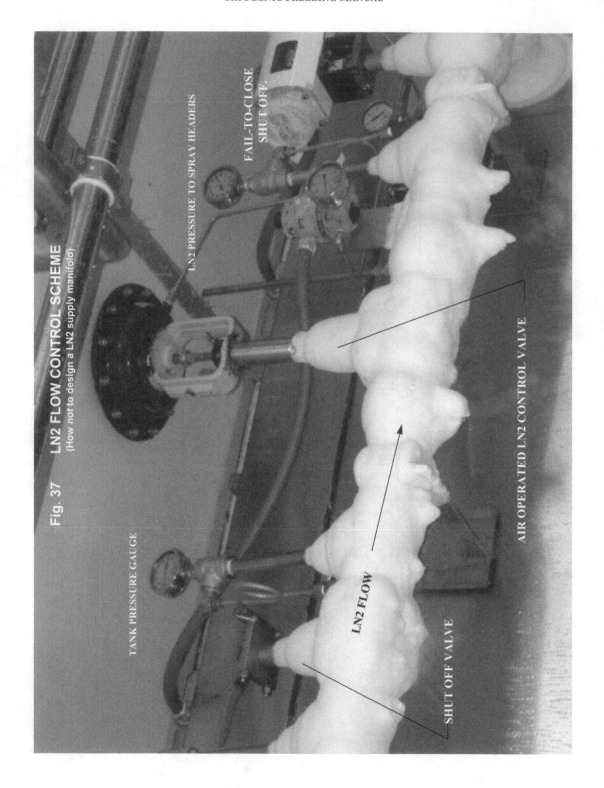

Fig. 37   LN2 FLOW CONTROL SCHEME
(How *not* to design a LN2 supply manifold)

LN2 PRESSURE TO SPRAY HEADERS

FAIL-TO-CLOSE SHUT OFF.

TANK PRESSURE GAUGE

LN2 FLOW

SHUT OFF VALVE

AIR OPERATED LN2 CONTROL VALVE

10. EXHAUSTS

10.1 Cryogenic Freezer Exhaust Systems

The exhaust "system" of a cryogenic freezer maintains a safe working environment and in addition provides for a correct vapor balance in the freezer, without which it is impossible to use the LN2 efficiently. Operators of LN2 freezers must be fully aware of the importance of this component of the freezer. The product freezing cost depends on it.

Cryogenic freezers use Liquid Nitrogen (LN2) as an *expandable* refrigerant. The liquid refrigerant LN2 is injected into the freezer and goes through a phase change to exchange its energy with a food item of a higher temperature. During this phase change, large amounts of Gaseous Nitrogen (GN2) are formed in the freezer and must be vented or exhausted to the exterior of the building to keep the processing environment safe.

Cryogenic freezer manufacturers have designed exhaust plenums sufficiently sized as to withdraw all nitrogen vapors from the freezer allowing it to run practically uninterrupted within the nitrogen concentration safety limits as directed by OSHA for occupied spaces. More details are available in Section 12 "Safety Guidelines".

The temperatures of the exhaust vapors that exit a cryogenic freezer are in most cases well below the freezing point of water. Since the exhaust vapors are saturated with water vapor, ice formation in the entire exhaust system will occur. This ice, at first in the form of snow, will eventually block the ducting and blower-housing preventing a proper nitrogen vapor exhaust. To prevent this, warm air is mixed with the cold nitrogen vapors to keep the plenum, ducting and blower ice free. Figure 40 is an example of a system used for extremely humid processing environments. In addition, be aware that;

To maintain a safe working environment the exhaust blower must function unobstructed.

10.2 Components of the EXHAUST System

10.2.1 The "Spray Counter Flow" Tunnels
One exhaust plenum at the feed end of the tunnel is sufficient. Preferably with a design that allows suction from above and below the belt. This is a highly effective method and will allow the

tunnel to operate for extended periods without icing up the exhaust passages.

10.2.2 The Exhaust Duct

The exhaust duct is the section that connects the tunnel's exhaust plenum to the exhaust blower, which is installed on an outside wall, on the roof of the building or directly on the freezer.

Never install butterfly dampers in any exhaust duct! They will certainly within a short time form ice blockages and prevent proper ventilation.

Do not undersize the duct diameter. Here, bigger is better. Design the duct with as straight a line as possible using only long radius elbows. The preferred material is Stainless Steel. Provide inspection and cleaning ports at the appropriate locations to maintain a sanitary freezing operation.

10.2.3 The Exhaust Blower

A blower used to exhaust very cold vapors laden with moisture is specifically designed not to ice up and is equipped with a "Self cleaning" impeller. Different manufacturers use different approaches to prevent the impeller from collecting ice and preventing proper ventilation. *A typical blower used to move warm air cannot be used for cryogenic exhaust applications.* Blowers for cryogenic applications have an impeller with not more than 4 to 8 blades casted or welded on a circular disc which is mounted on the drive shaft. All components are made of either aluminum or Stainless Steel. Exhaust blowers can be mounted directly on the tunnel as illustrated in Fig. 39. The blower is completely integrated to the controls of the tunnel and is equipped with a frequency controller to regulate the suction.

10.3   EXHAUST BLOWER SIZING

10.3.1   For a "Spray Counter Flow" LN2 Tunnel

The cold nitrogen vapors are only extracted from one end of the freezer and must be diluted with warmer ambient air to form an exhaust vapor with a temperature above 32°F to prevent the formation of ice. To warm up the GN2 vapors, warmer ambient air is mixed with it to get a combined vapor temperature above 32°F. How much warmer air is required depends on the temperature and the humidity of that air. Accurate calculations are possible to determine the volumes of warm air necessary to mix with the cold GN2. Here are guidelines derived from practical experiences used to size the blower with a freezer operating at a minimum set point temperature of -160°F.

10.3.2   For a "Convection Tunnel" and Spiral Freezer

The cold nitrogen vapors are extracted from both ends of the freezer and must be diluted with warm ambient air similar as shown for the spray tunnel. However, the GN2 vapors for the tunnel or spiral freezer should be mostly directed towards the feed end of the freezer with a 70% to 30% ratio (70% towards the product entry end and 30% toward the exit end) such as shown below.

Freezer set point temperature at -160°F

A.   Ambient air temperature ±70°F. RH 50%.
Freezer LN2 Consumption Rate: 1000   2000   3000   4000   5000   6000
(In Lbs/Hr.)

Total Exhaust Volume, CFM: 1000   2500   3500   5000   6000   7500

B.   Ambient air temperature ±45°F. RH 50% and > 50%.

Total Exhaust Volume, CFM: 1380   3000   4850   6450   8000   9700

Freezer set point temperature at -160°F

A.   Ambient air temperature ±70°F. RH 50%.
Freezer LN2 Consumption Rate; 1000   2000   3000

(In Lbs/Hr.)

Total Exhaust Volume, CFM: 700/300 1750/750 2450/1050

B.   Ambient air temperature ±45°F. RH 50% and > 50%.

Total Exhaust Volume, CFM: 966/414 2100/900 3395/1455

10.4   The critical vapor balance

In several of the previous chapters, we have seen how important it is to operate a cryogenic freezer, specifically a tunnel freezer, with a balanced exhaust. To balance a freezer means that it is operating with a positive internal pressure just high enough to keep the warmer exterior air out of the freezer. *The warmer air will add to the refrigeration load of the freezer and will increase the LN2 usage.*

A properly balanced "Convection Tunnel" freezer has a small amount of vapor spilling out of both ends of the freezer. A "Spray Counter Flow" freezer has a slightly heavier puff of vapors exiting from the feed end and a very small amount of vapor exiting from the exit end. The operator should be well instructed about this and should feel confidant to make the adjustments.

◊ It is extremely important that during idle periods of longer than 5 minutes, the vapor exhaust is reduced to the absolute minimum to economize on the LN2 usage. This is of course done in tandem with the LN2 supply to the freezer by increasing the tunnel set point temperature. These adjustments are automatically adjusted if the freezer is equipped with a PLC.

Fig. 39 EXHAUST DUCTING FOR COLD & HUMID ENVIRONMENTS

(Material: Stainless Steel or PVC)

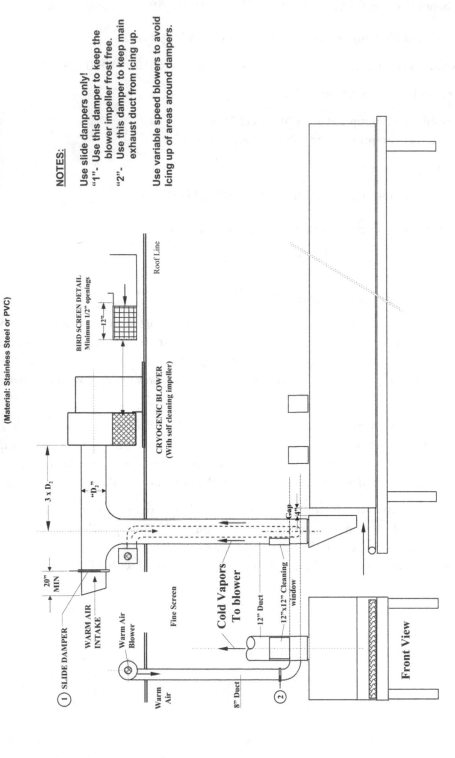

NOTES:

Use slide dampers only!

"1"- Use this damper to keep the blower impeller frost free.

"2"- Use this damper to keep main exhaust duct from icing up.

Use variable speed blowers to avoid icing up of areas around dampers.

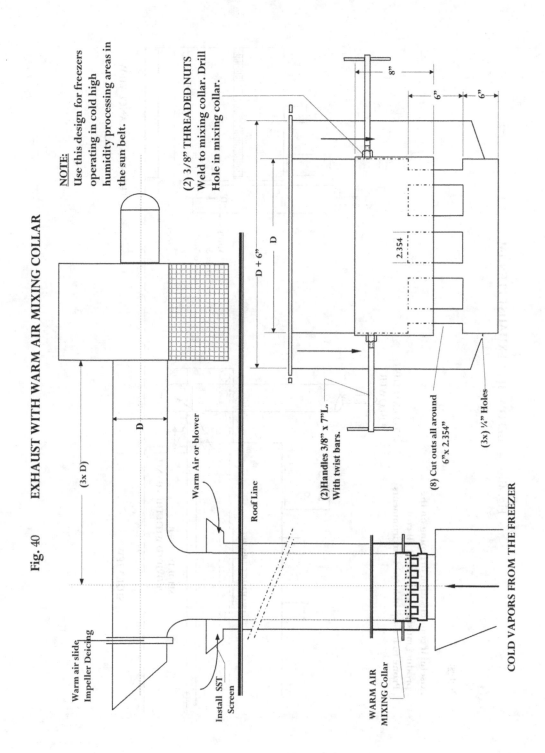

Fig. 40   EXHAUST WITH WARM AIR MIXING COLLAR

Fig. 41   BELOW BELT EXHAUST PLENUM

Notes:

. Useful if condensation dripping on the
  product must be avoided at all cost.
. Prone to ice-up in humid environments

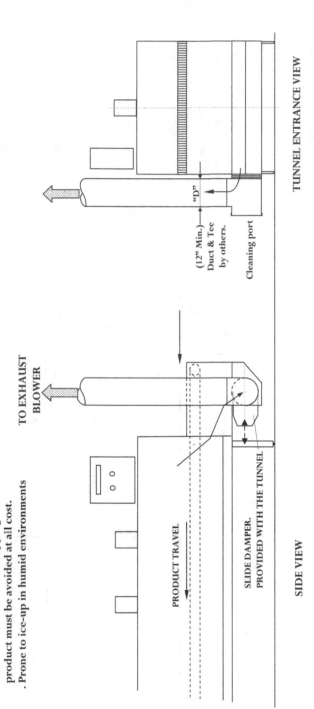

TO EXHAUST
BLOWER

PRODUCT TRAVEL

SLIDE DAMPER.
PROVIDED WITH THE TUNNEL

SIDE VIEW

(12" Min.)
Duct & Tee
by others.

"D"

Cleaning port

TUNNEL ENTRANCE VIEW

154

Figure 42
(Courtesy of CES,Inc. Cincinnati, OH.)

Figure 42 depicts one type of impeller used in cryogenic exhaust blowers.
To avoid icing up of the impeller, the spacing between the impeller blades are wide.
An other type is the Dayton Blower, it has only 4 blades on a circular back plate.

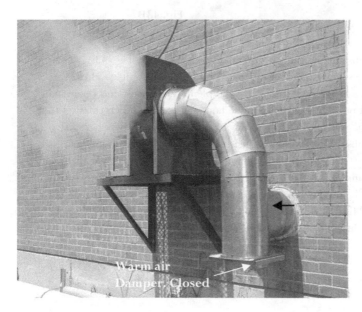

Figure 42A

A wall mounted Dayton blower. Shown with a standard 12" dia. galvanized duct. Notice the warm air slide damper. Its useful to prevent the blower impeller from icing up.

Exhaust Management

Fig. 43A   **Fig. 43B**

Figure 43 is a typical exhaust plenum arrangement of a LN2 tunnel. Fig. 43A has an almost good exhaust draw, slightly more draw would have been just right. However, notice that the slide damper in the duct just above the plenum is already full open and further adjustment was not possible. And within one hour the condition deteriorated to the point as shown in Fig. 43B. When the draw is hampered as shown here, two possibilities could be the cause. One, is the exhaust blower correctly sized, and two, is the duct obstructed with ice? After further examining this situation, it was clear that ice blockages had practically pinched off the exhaust duct in several places. The processing room temperature was 50°F and the humidity was at times as high as 85%. The tunnel was freezing raw marinated fish fillets with a tunnel set point temperature of -120°F. The problem was solved by installing a forced warm air make up as shown in Fig. 39

Fig. 44   Excessive nitrogen vapors at the tunnel exit. Increase the exhaust suction and/or increase the speed of the Scroll Fan.

Fig. 45   **Exhaust Blower mounted on the freezer**
(Courtesy CES,Inc.)

11. WHY IS THE FREEZER USING TOO MUCH LN2?

Here are several reasons,

11.1   The LN2 Tank Pressure is Too High

LN2 consumption = $\dfrac{\text{BTUs to be removed from the product}}{\textbf{BTUs available from the LN2}}$ = Lb.LN2/Lb.of prod.

In the above equation we see that the denominator energy from the LN2 (BTUs available from LN2) is responsible for the consumption to increase or decrease.

The LN2 tank pressure is directly linked to the quantity of cryogen used in the freezer. A high tank pressure will increase the flash-off rate of the nitrogen when it is injected in the freezer causing an imbalance of the internal freezer pressure, which will increase the rate of exhaust. A high tank pressure, which can be any pressure above one PSIG, is usually accompanied with a too low exhaust temperature. Exhaust temperatures, which are much lower than about 20 to 50 degrees of the product equilibrated temperature, are considered contributors to a high consumption rate.

Table 2 shows how the total energy derived from LN2 is a function of the pressure in the storage tank *and* the exhaust vapor temperature of the freezer.

$$(Q_{Total} = H_{vap} + SH)$$

A high tank pressure reduces the energy available from the LN2; this is obvious from the table. However, this is not where it ends. A cryogenic freezer supplied with LN2 from a "high pressure" tank becomes an operator's nightmare. The nitrogen vapors flashing off in the freezer are producing a positive pressure, which makes it very difficult to maintain the correct exhaust temperature. *Remember, we must keep the exhaust temperature as close as possible to the product-equilibrated temperature.*

Remedy: Drop the LN2 tank pressure as low as possible. Adjust this only while in full production and about two psig per adjustment, using the constant pressure controller. It is the gas supplier's responsibility to supply the processor with such a device. (See Fig.21) When you reduced the tank pressure too far, you cannot keep the freezer at the required set point temperature. Now back off on the pressure by 1-psig increments until the freezer's temperature gets to its set point temperature. Wait about 5 minutes between adjustments!!!

◊ The required tank pressure for a specific freezing operation can be found in the original project analysis prepared by the equipment vendor or your gas supplier.

11.2 Poor belt loading

A very common practice is to run the freezer with a belt loading much less than specified in the original project analysis. See section 5 for more details.

Poor belt loading will affect the specific LN2 consumption as follows:

1.   **Not enough products on the belt.** Running the freezer with a sparsely loaded belt requires a longer operating time to satisfy the required production throughput. This is clearly seen in the equation:

Prod. Rate = $\dfrac{\text{Belt Loading} \times L \times 6}{\text{Dwell Time}}$

Where, Belt Loading = Product weight in pounds per linear feet of belt.
L = Length of the freezer's freezing zone.

2.   **Too many pauses between production runs.** The product flow into the tunnel is halted for one reason or another and the freezer's temperature is maintained at its original set point temperature. If during a typical day too many of these interruptions occur, the freezer is wasting LN2. Do not overlook this aspect of the process.

11.3 Over freezing the product

Before assigning a person responsible for checking the product final temperature, he or she must be instructed how to measure the product temperature correctly. The product final temperature is the EQUILIBRATED temperature of the product after it exits the freezer. The correct procedure is to take about 5 pieces of the product and stack all five peaces tightly in a stainless steel insulated "Dewar" or similar device. Now insert a calibrated digital thermometer about halve way into the product mass. Hold it there until the temperature reaches its lowest point and begins to climb. This lowest temperature is the product's final equilibrated temperature. It can often take about 5 minutes before the procedure is finished. The equilibration phase takes place in the boxes the product is shipped in.

11.4 Exhaust not properly adjusted

Do not draw too much vapors from the freezer. Adjust the suction to a point where a small amount of vapors is escaping from the ends of the freezer. Observe and adjust this only after the freezer has been running with a *full* load of product for at least 10 minutes. It is common for the freezer, tunnels or spirals, to have a thermometer mounted in the exhaust duct just above the belt. It becomes obvious when too much of a suction is used too expel the exhaust vapors; the temperature will be much lower as recommended.

11.5 Not all turbulence fans are running

Check periodically if all the fans are running. Simply put the palm of your hand on the side of the motor housing to check this when the freezer has been running for a while. The heat transfer coefficient of the tunnel has been determined with a specific air velocity in the tunnel, which can only be achieved when all the turbulence fans are running.

The air speed down draft of each fan should be around 1500 ft/min.

Most important: Make sure that the fan blades have enough clearance between the blade and the interior tunnel top. The *minimum* dimension is approximately 1/3 of the fan's diameter. Each manufacturer has a specific requirement for this dimension.

11.6 Spray nozzles too large

Do not overlook this. The LN2 header of a cryogenic freezer is designed with just the right number and size of nozzle to introduce a specific amount of LN2 in the freezer. Each nozzle should spray *fine* droplets of liquid in the freezer and in a spray counter flow tunnel on the product. Too large of a hole in the nozzles will spill LN2 in the tunnel at a rate too high to vaporize properly and could run of the product onto the bottom of the freezer. Do not tamper with the nozzles until you have consulted with the manufacturer. Each equipment manufacturer has its own design criteria for sizing the size and number of nozzles required to introduce enough energy into the freezer. As a rule of thumb, *each* nozzle should have a capacity of approximately 0.2 GPM at 20 PSIG and spray *directly downwards* for a spray counter flow tunnel. For a convection tunnel, direct the nozzles horizontally into the circulating fans down draft. The "Spraying Systems" data book is an excellent source for this calculation.

11.7 The LN2 tank was not filled properly

This is a topic not often addressed. Chapter 8, paragraph 8.1.6 addresses this topic sufficiently. Make it your business to know how the filling procedure is done. It will affect the freezing cost and ad up dramatically when the tank is filled more than once a week.

Of course, none of the above applies if the freezer is not properly sized.

11.8 Guaranteed ways to fail your freezing operation

In this manual, failing the freezing operation only points to a high freezing cost. This is certain if we;

1. Totally ignore the LN2 tank. Never observe how the tank is filled, and never question the tank's vacuum condition.
   Remedy: We should at all times know the condition of the vacuum in the annular space between the inner and outer vessel. Request a written report at least once every 2 years if the tank is leased from the gas supplier. If the LN2 tank is your own property, have this checked. See chapter 8.1 for details. During the tank fill, make sure that the required tank pressure remains steady without a constant vapor release from the tank's vent line. Avoided this by using the "top fill" only during the last one third of the fill process.

2.   Use polyurethane insulated piping with inadequate insulation. Or the insulation is more than 7 years old. Or worst, leave sections of the piping bare with no insulation at all.
Remedy: Use PU insulated pipe with a minimum of 4-inch thick insulation. **Replace** PU insulated piping older than 7 years. Best solution, use Vacuum Jacketed pipe. It will last for at least 20 years. At that time, a new vacuum can be drawn to get an as-new piping system.

3.   Ignore the fact that the tank pressure is well above 20 psig for a vertical tank and 10 psig for a horizontal tank, to get the freezer to the desired set point temperature.
Remedy: Examine if the LN2 piping has the required diameter. If for instance the pipe size is too small for a certain flow requirement, the tank pressure must be increased to get the required flow rate. The only remedy here is to install the correct pipe size.

4.   Ignore a high tank pressure. The cause can be a simple adjustment of the pressure regulator. This simple adjustment can be responsible for a 5% to 10% increase in LN2 usage.

5.   Do not pay any attention to the exhaust management of the freezer. The exhaust temperature must be within the temperature tolerances as mentioned in chapter 10.
Remedy: Correct this condition by adjusting the blower suction or by moving the duct dampers to the correct position until the exhaust is balanced.

6.   Measure the required product temperature immediately after it exits the freezer. Remember that only a measurement known as the equilibrated temperature will give you the correct final product temperature. This temperature is usually reached after about 5 minutes depending on the mass of the item. The thicker the product item, the longer it takes to reach the equilibrated temperature.
Remedy: Instruct the QC person properly.

7.   Operate the freezer with a poor belt loading. In other words, the freezer's conveyor belt has large section of the belt surface running through the freezer without product on it.
Remedy: Examine how the product is fed onto the freezer belt and make the necessary corrections.

8.   Operate the freezer with one or several circulating fans not running. This is a recipe for disaster, which will certainly result in excessive LN2 usage.
Remedy: Correct the problem first. This is often caused by a burnt out motor or just a simple tripped starter switch. Call an electrician.

9.   Run the freezer with a set point temperature, several degrees lower than recommended in the manual.
Remedy: Look up what is recommended and adhere to this temperature.

10.   Start the freezer when it is still dripping wet from the cleaning process. This additional heat load from water will require extra LN2 to chill down the freezer.
Remedy: Turn on the freezer about 1/2 hour prior to the arrival of the product *without* the LN2 on but run the exhaust blower and all circulating fans at full power with a running belt. Completely open the dampers in the exhaust duct.

11.   Start the LN2 injection much too early. It is not necessary to begin the LN2 injection

more than 20 minutes prior to the arrival of the product.

Remedy: Instruct the operator properly. If the tunnel does not reach the required set point temperature within 20 minutes, not enough LN2 is injected into the tunnel. Examine if the nozzles are sized correctly or are plugged up.

12. Operate the freezer without a PLC unit. A PLC can be pre-programmed to do all of the above steps without human interface and will guarantee a LN2 consumption as initially calculated.

    Remedy: The freezer is a major part of your cost basis. Do not approach it on the cheap. Order the freezer with a pre-programmed PLC.

13. Running too much product through the freezer.

    Remedy: Order a bigger freezer. Tunnels can, in most cases, be extended in length to process more products. Contact the manufacturer to investigate if this option is possible.

12. SAFETY – NITROGEN

12.1 Nitrogen (N_2)

12.2 Oxygen Deficient Atmospheres

12.3 The two forms of nitrogen
12.3.1 Liquid Nitrogen (LN2)
12.3.2 Nitrogen Vapor (GN2)

12.4 General Safety Guidelines

12.5 Possible danger
12.5.1 Oxygen Deficiency due to Air Displacement
12.5.2 Cold Burns
12.5.3 High Pressure Injury
12.5.4 Oxygen Monitors

Figure 46   Typical Oxygen Monitor Assembly

12.1 Nitrogen (N$_2$)

Chapter 1, paragraph 1.2.1 describes what nitrogen as an element is and the specifics of its properties. A boiling point of -320°F is of course why this element, in its liquid form, is used as a refrigerant. The safety aspects of this refrigerant, as specified by the Compressed Gas Association, Inc., Arlington, VA 22202 in the "Safety Bulletin" number SB-2 – 1983 are to be adhered to if a safe working environment is required.

12.2 Oxygen Deficient Atmospheres

Air we breathe contains 21% of oxygen by volume, which is required to sustain life. The dilution or depletion of this oxygen in air by displacement with an inert gas, such as nitrogen, is a hazard to personnel. Below are the oxygen limits required for an area where personnel is present as prescribed by the Compressed Gas Association.

Oxygen content effects and symptoms at atmospheric pressure.

(Percentage By volume)

15 – 19%	Decreased ability to work strenuously. May impair coordination and may induce early symptoms in persons with coronary, pulmonary, or circulatory problems. **(Minimum allowed)**
12 – 14%	Respiration increases in exertion, pulse up, impaired coordination, perception, and judgment. (This level should *never* be reached)
10 – 12%	Respiration further increases in rate and depth, poor judgment, blue lips.
8 – 10%	Mental failure, fainting, unconsciousness, ashen face, blueness of lips, nausea, and vomiting.
6 – 8%	8 minutes, 100% fatal; 6 minutes, 50% fatal; 4-5 minutes, recovery with treatment.
4 – 6%	Coma in 40 seconds, convulsions, respiration ceases, death.

NOTE: Exposure to atmospheres containing 12% or less oxygen will bring about unconsciousness without warning so quickly that the individual cannot help or protect him/herself

12.3   The two forms of nitrogen

12.3.1   Liquid Nitrogen (LN2)

LN2 is used as an expandable refrigerant in cryogenic freezers to freeze foodstuffs and in the process changes from a liquid state to a vapor state. This change of state takes place at approximately -306°F and 20 psig LN2 supply pressure.

12.3.2   Nitrogen Vapor (GN2)

Nitrogen vapors are still at a very low temperature when due to a pressure differential will escape into the processing room when it is not properly vented. These cold vapors are heavier than air and thus will settle to the floor and eventually displace the oxygen in the area resulting in a hazardous working environment. Also see chapter 10.

12.4   General Safety Guidelines

Operating and maintenance personnel should at all times perform their duties with safety in mind. When working on or around the nitrogen freezers, observe the following guidelines:

1.   Be certain that the area around the freezer has adequate ventilation to prevent nitrogen vapors from accumulating. Specifically near the floor.

2.   At all times during the freezer operation, the exhaust blower must be operating.

3.   Wear protective clothing and gloves when working with cold freezer components that were exposed to LN2.

4.   Avoid working near the floor when the freezer is in operation. Cold nitrogen vapors will concentrate near the floor.

5.   Wear an oxygen mask with an independent oxygen source when entering an area of known high nitrogen levels.

6.   Emergency rescue personnel with breathing gear should stand by whenever someone enters a high nitrogen concentration area.

7.   Always provide pressure vents (Safety relief devices) when LN2 can be trapped between two valves in a supply line. Unless this pressure can bleed back into the LN2 tank or out of a safety valve, the line will rupture.

8.   Safeties should be installed to prevent harming personnel when it is relieving pressure.

9.   ***Provide enough make up air to the processing room to avoid a negative pressure from developing. The exhaust blower is removing not only the nitrogen vapors escaping from the freezer, but some of the room air as well.***

10.   Consult a knowledgeable engineer when installing LN2 piping and/or valves to ensure proper safety valve sizing and ventilation requirements.

12.5 Possible danger

Inform all personnel about the potential hazards of LN2 used in freezing equipment. Such as:

12.5.1 Oxygen Deficiency due to Air Displacement

If nitrogen were allowed to flow continuously into a sealed room, nitrogen vapors will build up to a level where it displaces the oxygen necessary for breathing. Personnel entering this room would eventually become dizzy, then unconscious and eventually suffocate.

Operators of LN2 freezing equipment are responsible to provide a safe working environment and that all personnel working in the area are informed about potential hazards related to the use of LN2 or any other equipment utilizing nitrogen that are able to displace oxygen.

Avoid buildup of excessive nitrogen vapor concentrations by taking the following precautionary measures:

- Provide continuous monitoring of the oxygen concentrations at the most critical locations (near the entrance and exit of the freezer) in a processing room if the area is small and is not sufficiently ventilated.

- Ensure proper mechanical ventilation of the work areas to prevent excessive nitrogen vapor build up.

- Be absolutely sure that the spillage of nitrogen vapors from the freezer is kept to an absolute minimum.

- Check if the exhaust fan is turned on and operating effectively by examining the vapor flow exiting the freezer's inlet and outlet.

- Make sure that the exhaust is properly vented to an outdoor location, properly sized and not obstructed with ice during long production runs.

12.5.2 Cold Burns

Although all freezing equipment is designed to completely enclose and insulate the cold surfaces, accidental damage to a LN2 line in the freezer could occur and may lead to LN2 running out of the freezer. Contact with LN2 or a jet of cold liquid nitrogen will cause severe burn like injuries.

Also, the surface of a product leaving the freezer has a subzero temperature and should not be handled without protection. This applies to the internal surfaces of a just opened freezer as well.

> *WARNING: Do not touch Liquid Nitrogen, frozen equipment, piping or product with bare hands!*
>
> *Never play with LN2.*
>
> *Use eye protection when working with LN2.*

12.5.3 High Pressure Injury

> *LN2 pipelines should be regarded as high pressure vessels such as with compressed air or steam.*

Be careful not to keep LN2 trapped in a pipeline. As the liquid warms, it expands while changing into a gas causing the pressure to increase.

12.5.4 Oxygen Monitors

To ensure a safe working environment, an oxygen monitor should be installed near the freezer. For proper measurement of the oxygen concentrations in the processing room, the sensor must be mounted approximately 42 inches above the floor in *still air*.

If for any reason the nitrogen level should build up, the sensor will detect a decrease in the amount of oxygen in the processing room, sound an alarm and shut down the LN2 supply to the freezer. This control feature should always be included in the freezer's control system when the processing room is small and the supply of make-up air is questionable. If the processing room is fairly large, approximately (100' x 200' x 20' high) and sufficient fresh air enters the processing room, a portable oxygen monitor which the floor supervisor can carry is sufficient.

Fig. 46   Typical Oxygen Monitor assembly

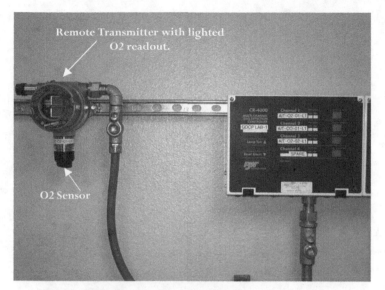

4 Station Monitor

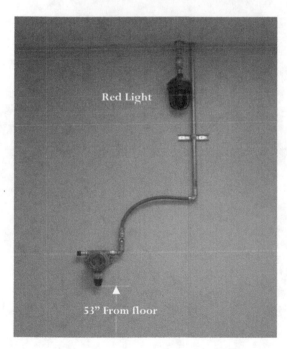

Shown here is a 4 station Monitor with an oxygen transmitter assembly mounted next to it.

In addition, a second transmitter assembly with an alarm light has been installed in an adjacent processing room.

Install the O2 sensor 53" of the floor in <u>still</u> air.

(Courtesy of BW Technologies)

13. PHYSICAL PROPERTIES OF VARIOUS FOOD STUFFS

PROPERTIES FOR FOODS
Table 9

FRUITS & VEGETABLES

COMMODITY	AVERAGE FR.POINT F	WATER CONTENT %	SPECIFIC HEAT BTU/LB./F ABOVE FR.	BELOW FR.	LATENT HEAT BTU/LB. (Calc.)
Apples	28.20	84.10	0.87	0.45	121
Apricots	29.60	85.40	0.88	0.46	122
Artichokes					
Globe	29.60	83.70	0.87	0.45	120
Jerusalem	27.50	79.50	0.83	0.44	114
Asparagus	30.40	93.00	0.94	0.48	134
Avocados	30.00	65.40	0.72	0.40	94
Bananas	29.60	74.80	0.80	0.42	108
Beans					
Green or Snap	30.20	88.90	0.91	0.47	128
Lima	30.80	66.50	0.73	0.40	94
Dried	-	12.50	0.30	0.24	18
Beets (Topped)	29.20	87.60	0.90	0.46	126
Blackberries	29.40	84.80	0.88	0.46	122
Blueberries	28.60	82.30	0.86	0.45	118
Broccoli	30.30	89.90	0.92	0.47	130
Brussel Sprouts	30.20	84.90	0.88	0.46	122
Cabbage	30.50	92.40	0.94	0.47	132
Carrots (Topped)	28.80	88.20	0.9	0.46	126
Cauliflower	30.20	91.70	0.93	0.47	132
Celery	30.90	93.70	0.95	0.48	135
Cherries	27.70	83.00	0.87	0.45	120
Corn (Sweet)	30.80	73.90	0.79	0.42	106
Dried	-	10.50	0.28	0.23	15
Cranberries	30.00	87.40	0.9	0.46	124
Cucumbers	30.50	96.10	0.97	0.49	137
Currants	30.20	84.70	0.88	0.45	120
Dewberries	29.20	-	-	-	-
Dates (Fresh)	27.10	78.00	0.82	0.43	112
Eggplant	30.40	92.70	0.94	0.48	132
Endive(Escarole)	31.10	93.30	0.94	0.48	132
Figs (Fresh)	27.10	78.00	0.82	0.43	112
Dried	-	24.00	0.39	0.27	34
Garlic (Dry)	28.00	74.20	0.79	0.42	106
Gooseberries	30.00	88.90	0.9	0.46	126
Grapefruit	28.40	88.80	0.91	0.46	126
Grapes					
European	27.10	81.60	0.86	0.44	116
American	29.40	81.90	0.86	0.44	116
Horseradish	26.40	73.40	0.78	0.42	104
Huckelberries	28.60	82.30	0.86	0.45	119

PROPERTIES FOR FOODS
Table 10

FRUITS & VEGETABLES

COMMODITY	AVERAGE FR.POINT F	WATER CONTENT %	SPECIFIC HEAT BTU/LB./F ABOVE FR.	BELOW FR.	LATENT HEAT BTU/LB. (Calc.)
Kale	30.70	86.60	0.89	0.46	124
Kohlrabi	30.00	90.10	0.92	0.47	128
Leeks	30.40	88.20	0.90	0.46	126
Lemons	29.00	89.30	0.92	0.46	127
Limes	28.20	86.00	0.89	0.46	122
Lettuce	31.20	94.80	0.96	0.48	136
Loganberries	29.50	82.90	0.86	0.45	118
Mangoes	29.40	81.40	0.85	0.44	117
Melons					
Cantalope & Persian	29.90	92.70	0.94	0.48	132
Honey Dew & Honey Ball	29.80	92.60	0.94	0.48	132
Casaba	29.90	92.70	0.94	0.48	132
Watermelon	30.60	92.10	0.97	0.48	132
Mushrooms (Fresh)	30.00	91.10	0.93	0.47	130
Nectarines	29.00	82.90	0.90	0.49	119
Okra	28.60	89.80	0.92	0.46	128
Olives (Fresh)	28.50	75.20	0.80	0.42	108
Onions	30.10	87.50	0.90	0.46	124
Oranges	28.00	87.20	0.90	0.46	124
Papayas	30.10	90.80	0.82	0.47	130
Parsnips	29.80	78.60	0.84	0.46	112
Peaches	29.60	86.90	0.90	0.46	124
Pears	27.70	82.70	0.86	0.45	118
Peas (Green)	30.10	74.30	0.79	0.42	106
Dried	-	9.50	0.28	0.22	14
Peppers (Sweet)	30.50	92.40	0.94	0.47	132
Chili (dry)	30.90	12.00	0.30	0.24	17
Persimmons	27.50	78.20	0.84	0.43	112
Pineapples					
Mature-Green	29.10	-	-	-	-
Ripe	29.70	85.30	0.88	0.45	122
Plums	28.70	85.70	0.88	0.45	123
Pomegranates	26.50	77.00	0.87	0.48	112
Potatoes					
Early crop	30.00	-	-	-	-
Late crop	29.80	77.80	0.82	0.43	111
Sweet	29.20	68.50	0.75	0.40	97
Prunes - Fresh	28.70	85.70	0.88	0.45	12
Pumpkins	29.90	90.50	0.92	0.47	123
Quinces	28.10	85.30	0.88	0.45	130
Radishes	30.10	93.60	0.95	0.48	122

PROPERTIES FOR FOODS
Table 11

FRUITS & VEGETABLES

COMMODITY	AVERAGE FR.POINT	WATER CONTENT	SPECIFIC HEAT BTU/LB./F		LATENT HEAT BTU/LB.
	F	%	ABOVE FR.	BELOW FR.	(Calc.)
Raspberries					
Black	29.40	80.60	0.84	0.44	122
Red	30.30	84.10	0.87	0.45	121
Rhubarb	29.90	94.90	0.96	0.48	134
Rutabagas	29.70	89.10	0.91	0.47	127
Salsify	29.60	79.10	0.83	0.44	113
Spinach	31.30	92.70	0.94	0.48	132
Sauerkraut	26.00	89.00	0.92	0.47	128
Squash					
Acorn	30.00	-	-	-	-
Summer	30.40	95.00	0.96	0.48	135
Winter	29.80	88.60	0.91	0.47	127
Strawberries	30.20	89.90	0.92	0.47	129
Tangerines	29.50	87.30	0.90	0.46	125
Tomatoes					
Mature-Green	30.40	94.70	0.90	0.48	134
Ripe	30.40	94.10	0.95	0.48	134
Turnips	29.80	90.90	0.93	0.47	130
Vegetables(Mixid)	30.00	90.00	0.90	0.45	130

PROPERTIES FOR FOODS
Table 12

MEATS

COMMODITY	AVERAGE FR.POINT F	WATER CONTENT %	SPECIFIC HEAT BTU/LB./F ABOVE FR.	SPECIFIC HEAT BTU/LB./F BELOW FR.	LATENT HEAT BTU/LB. (Calc.)
Bacon - Cured					
Raw Slab	-	20.0	0.60	0.36	29
Beef					
Lean Sirloin Steak					
Raw	29.00	71.80	0.82	0.45	104
Cooked	-	58.70	0.70	0.43	85
Hamburger - Lean					
Raw	29.00	68.30	0.80	0.45	98
Cooked	-	60.00	0.75	0.43	86
Hamburger - Regular					
Raw	28.00	60.20	0.77	0.43	87
Cooked	-	54.20	0.72	0.42	78
Liver - Beef	29.00	69.70	0.80	0.45	100
Chicken					
Chicken Fryers					
Flesh & Skin-Raw	27.00	75.40	0.84	0.45	109
Fried	-	53.50	0.71	0.42	77
Lamb					
Loin Lean					
Raw	29.00	71.80	0.82	0.45	104
Cooked	-	61.30	0.75	0.44	88
Ground-Raw	29.00	68.00	0.8	0.44	98
Pork					
Loin - Lean	28.00	67.50	0.80	0.44	97
Loin - Med. Fat					
Total Edible - Raw	28.00	57.20	0.75	0.43	82
Roasted	-	45.80	0.69	0.41	66
Separable Lean					
Raw	28.00	67.50	0.79	0.44	97
Roasted	-	55.00	0.72	0.43	79
Spareribs-Med.Fat					
Raw	28.00	51.80	0.73	0.42	75
Braised	-	39.70	0.66	0.40	58
Ground Pork	28.00	66.00	0.81	0.44	95
Sausage - Fresh	26.00	38.00	0.67	0.39	55
Veal					
Cutlet - Raw	29.00	71.00	0.84	0.45	102

PROPERTIES FOR FOODS
Table 13

SEAFOODS

COMMODITY	AVERAGE FR.POINT F	WATER CONTENT %	SPECIFIC HEAT BTU/LB./F ABOVE FR.	BELOW FR.	LATENT HEAT BTU/LB. (Calc.)
Fish- Whole					
Bass					
Black and Sea	28.00	79.30	0.84	0.44	114
Striped	28.00	77.70	0.82	0.43	112
White	28.00	78.80	0.83	0.44	113
Bluefish	28.00	75.40	0.80	0.43	108
Catfish - Fr Watr	28.00	78.00	0.82	0.43	112
Cod	28.00	78.00	0.82	0.43	112
Haddock	28.00	78.00	0.82	0.43	112
Halibut	28.00	75.00	0.80	0.43	108
Manhaden	28.00	62.00	0.70	0.38	89
Salmon	28.00	64.00	0.71	0.39	92
Fish Fillets					
Cod	28.00	80.00	0.84	0.44	115
Haddock	28.00	80.00	0.84	0.44	115
Hake	28.00	82.00	0.86	0.45	118
Herring					
Raw	28.00	70.00	0.76	0.41	100
Smoked	-	64.00	0.71	0.39	92
Pickled (Bismark)	-	59.40	0.76	0.38	86
Mackerel	28.00	57.00	0.66	0.37	82
Ocean Perch	28.00	80.00	0.84	0.44	115
Pollock	28.00	79.00	0.83	0.44	113
Whiting	28.00	82.00	0.86	0.45	118
SHELLFISH					
American Lobster	28.00	79.00	0.83	0.44	113
Clams					
Meat Only	28.00	81.70	0.87	0.45	117
Meat & Licquor	28.00	87.00	0.9	0.46	125
Oysters					
Meat Only	28.00	81.70	0.87	0.45	117
Meat & Licquor	28.00	87.00	0.9	0.46	125
Scallop Meat	28.00	79.80	0.84	0.44	115
Shrimp	28.00	83.00	0.86	0.45	119

PROPERTIES FOR FOODS
Table 14

DAIRY PRODUCTS

COMMODITY	AVERAGE FR.POINT F	WATER CONTENT %	SPECIFIC HEAT BTU/LB./F		LATENT HEAT BTU/LB. (Calc.)
			ABOVE FR.	BELOW FR.	
Butter					
Unsalted	32.00	15.8	0.57	0.35	23
Salted 2%	15.80	15.80	0.57	0.35	23
Salted	-3.60	15.80	0.57	0.35	23
Cheese					
American Process	16.60	40.00	0.65	0.39	58
Cheddar	19.60	38.80	0.62	0.32	56
Cottage	29.80	78.70	0.86	0.46	113
Swiss - Domestic	15.00	40.00	0.64	0.39	58
Eggs - Shell	28.00	67.00	0.74	0.40	96
Ice Cream	22.29	58-66	0.80	0.45	96
Milk - Whole	31.00	87.50	0.93	0.49	124

PROPERTIES FOR FOODS
Table 15

TYPICAL PREPARED FOODS

COMMODITY	AVERAGE FR.POINT F	WATER CONTENT %	SPECIFIC HEAT BTU/LB./F		LATENT HEAT BTU/LB. (Calc.)
			ABOVE FR.	BELOW FR.	
Prepared Entrees					
Ham in Raisin Sauce		68.20	0.79	0.45	98
Macaroni & Cheese		78.10	0.85	0.46	112
Meat Balls in Tomato Sauce		75.70	0.84	0.46	110
Sliced Beef W. Gravy		73.70	0.83	0.46	107
SlicedTurkey w. Gravy		78.00	0.85	0.46	112
Breaded Flounder		64.00	0.76	0.44	92
Breaded Shrimp		63.00	0.76	0.44	91
Chicken ala King		68.20	0.79	0.45	98
Chili con carne w. beans		72.40	0.81	0.45	104
Chop suey w. meat (cooked)		75.40	0.84	0.46	108
Black Sea Bass Stuffed - Baked		52.90	0.69	0.43	76
Prepared Vegatables					
Asparagus cuts & tips		85.20	0.90	0.48	123
French fried onion rings		61.70	0.75	0.44	89
Green beens w. Mushroom sauce.		82.70	0.89	0.48	119
Potatoes au Gratin		72.00	0.81	0.45	104
Peas & Onions in sauce		80.30	0.88	0.47	115
PIZZA					
Cheese Topping-Baked.		48.30	0.68	0.42	62
Sausage Topping Baked.		50.60	0.68	0.42	74

14. A FINAL WORD

Now that we have gone through all the preceding chapters, one should realize that a cryogenic freezer is not some mysterious piece of machinery. In fact, it is considerably less complicated than many other pieces of equipment in the food processing plant. But for one reason or another, plant engineers seem to have a cavalierish attitude towards the cryogenic freezer. I have never understood this. Specifically, since it will affect the production cost directly if not designed properly. In other words, the entire freezing system should have been given the utmost attention during the planning stages.

> *Do not cut corners! Install the best there is. You will safe by using the latest technically up-to-date freezers, piping and cryogenic receivers.*

Although the following topic is previously discussed, its importance warrants to repeat this again.

14.1 Freezing With Liquid Nitrogen (LN2)

At a certain juncture during the planning stages, management has made the decision to select a LN2 system to freeze their product.

This decision was made based on a thorough understanding of the different options and why for this processor cryo-freezing will provide the best available solution. Now we have reached, in my opinion, a critical point in the planning sequence. Whom do you contact first to get honest relevant information? Do you know which questions to ask to weed out individuals our companies that are not working for your best interest? It may be beneficial to attain the services of a consultant to assist you in designing a freezing system best suited for your needs. I hope that this manual can aid you to make this decision.

14.1.1   The Cryogen Supplier
In chapter 8, we discussed how important the quality of the LN2 is when used in a freezing tunnel or any other type of cryogenic freezer. It should be in the super critical (without vapor) state when delivered to minimize the blow-off losses during transfer from the tanker truck to the receiver. Hence, select the supplier, which has an air separation plant the closest to your plant. Make sure that the liquid transport arrives at the plant with a pressure around 20 PSIG and is not venting vapor. When it is venting from the transport tank, it will be venting from the LN2 receiver as well.

Consequently, all of the venting losses in the LN2 receiver occur after the billing meter, which means that you will be billed for these losses.

Do not forget, you must insist that the transport's tank pressure is noted on every delivery ticket. In addition, the customer's tank pressure before and after the filling process must be noted on the delivery ticket as well. This data is necessary for you to calculate how much LN2 is used to freeze your product.

14.1.2   Is The Supplier Competent?

Question the supplier about their knowledge of the basics of cryogenic engineering. Are they able to support your engineering needs when the need arises? Use this manual as your guide.

They must be able to assist you to determine if you are freezing efficiently. Quantity of LN2 used to freeze the product.

14.2   The LN2 Freezer

This is where the rubber meets the road. Both the LN2 supplier and the equipment manufacturer must be consulted in the selection process. If Section "A" is well studied, one should have a basic understanding how a freezer should be designed. What is a good freezer? Here are some pointers:

1. The design must have the ability to integrate all vital components to be controlled by a PLC for best LN2 consumption

2. The freezer must be part of a system (LN2 tank, piping and freezer) that allows the operator to determine easily how much LN2 is used to freeze a product

3. Must have a simple LN2 supply system. See Section 9.

4. With item 3 satisfied, the LN2 pressure should be able to be as low as 20 psig for a vertical tank and 10 psig for a horizontal tank

5. The Exhaust System must be designed so that the internal vapor pressure in the freezer can be regulated easily

6. A tunnel should have an internal scroll van to control the internal vapor flow for tunnels longer than 20 feet. See Fig. 24A

7. The circulating vans should have an air velocity of at least 1600 ft/min.

8. The freezer should be easily accessible for cleaning and maintenance

9. An adjustable slide or baffle down stream from the spray zone should be part of the design. This prevents the vapor to flow towards the exit opening. This is important! See Fig. 24A.

10. A splash plate should be installed under the LN2 spray zone. This prevents LIQUID nitrogen from reaching the bottom of the freezer. See Fig. 24A.

11. For tunnels longer than 30 feet, two temperature sensors should be installed for independent or joint sensing. One near the LN2 spray zone and one near the exit. See Section 9 for more details.

12. Use a wide mesh belt if the product allows it. This will satisfy the circulation requirements

of an efficient freezer.

14.3 The exhaust

As explained in Chapter 10, this is a part of the installation, which will ultimately determine how efficient the freezer, and how safe the processing room is. Only use an exhaust blower designed to function under extreme low temperatures. By all means, do not cut corners with this part of the freezer. Always install a blower with a variable speed controller for the proper exhaust management. When using a PLC as the logic for the freezer's control, a variable speed control is inevitable. Under no circumstance, use ducting less than 10 inches in diameter. A 12-inch duct is better. Only use long radius elbows where elbows are needed. When dampers are required in a duct assembly, slide dampers are better than butterfly dampers. Butterfly dampers cause snow and ice to block the vapor flow through the ducting within half an hour.

NOTE: The design of a well-engineered exhaust system is perhaps the most challenging part for the freezer's manufacturer. The shorter the tunnel, the more difficult it is. As mentioned previously, not anything shorter than 20 feet is economical from a freezing cost point of view.

14.4 The LN2 Tank or Receiver

Most likely, the cryogen supplier is the one that will determine the size of the receiver the customer will get. In Chapter 8, we have stressed the importance of the LN2 tank in the freezing system.

A strong statement but we have encountered such situations at more than one occasion. Just imagine a large producer of poultry products is using LN2 to freeze chicken breast from 40°F to -10°F in a "Spray Tunnel" with a set point temperature of -160°F and an exhaust temperature of 0°F. From available tables such as shown in Chapter 13, we can calculate that 137 BTUs per pound of chicken must be removed. The LN2 tank was a horizontal tank with a tank pressure of 56 PSIG, installed about 150 feet from the freezer. The LN2 line was a Polyurethane insulated copper pipe of sufficient diameter.

From Table 2 we see that at this pressure the maximum BTUs available per pound of nitrogen are 150.9 (With an exhaust temperature of 0°F) For our further evaluation we used a tunnel efficiency of 80%. We calculated that the LN2 consumption should have been, 137/ (0.8 x 150.9) = 1.13 Lb/Lb.

Instead, the average consumption was 2.5 Lb/Lb.

We began corrective measures by installing a horizontal LN2 tank on a 30 feet steel structure and were able to reduce the tank pressure to 10 PSIG. From Table 2 we see that at 10PSIG the available BTUs are now 161.7. About 10% more. In addition, we had to increase the LN2 header and control valve on the tunnel. The spray nozzles were increase in diameter by about 50% to accommodate sufficient flow at the lower tank pressure. The results were amazing. The LN2 consumption was measured to be about 1.2 Lb/Lb. In theory, the calculated consumption should have been 137/ (0.8 x 161.7) = 1.1 Lb/Lb. We noticed that the exhaust vapor temperature dropped to − 40°F with the made modifications. The scroll fan speed was reduced to get to the original 0°F.

Fortunately, when one begins to get into the design stages of the cryogenic freezing system, the LN2 tank is the first topic the gas supplier will address. Get involved! Do not rely solely on others to determine these critical segments of your freezing system. Your profitability will depend on it.

14.4.1   Tank Pressure Management

The LN2 tank should be able to function with a constant pressure and as low a pressure as the system allows. See Fig. 21 & 22 for the specifics. Do not rely for the tank pressure control on the diaphragm type regulators installed on all cryogenic tanks. They are not able to control pressures within 1 (ONE) PSI.

14.4.2   Tank Vacuum

Do not overlook this. Request in writing what the condition of the vacuum in the annular space is and when the reading was made. See section 8.1.2 of Paragraph 8 for more specifics. Request from your gas supplier data on the frequency of the tank's vacuum checks, and if necessary when the next vacuum draw is scheduled. Keep your own log of this data.

14.5   The LN2 conduit or piping

Not just some but all the installations we surveyed used very poorly insulated or wrongly sized piping. It may have something to do with the notion that LN2 behaves similarly as water in a piping system. Use Paragraph 8, as your guide. Make a very calculated decision if you should use vacuum jacketed or Polyurethane insulated piping. It is more than likely, that the ultimate LN2 consumption in the above calculation could have been 1.13 Lb/Lb. instead of 1.4 Lb/Lb or 23% less if the customer had used a vacuum-jacketed piping system.

Configure the piping layout as direct as possible. The distance between the LN2 tank and the freezers should be as short as possible. Eliminate as many elbows and tees as possible. A total equivalent pipe length of more than 100 feet should always be a vacuum-jacketed pipe. See section 8.3.2 of Paragraph 8.

If more than 1 drop is required, use the design as illustrated in Fig.17

14.6   Product storage room temperature

Most cryogenic freezing operations freeze the product in the LN2 tunnel after which the product is packaged, palletized, and stored in refrigerated storage rooms. The temperatures in these storage rooms are as low as − 20°F and as high as 0°F. Deteriorative processes in the food item are temperature dependent. Of course, the original quality of the product plays a role in this as well. As a result, the science of product preservation by means of low temperatures is well understood by the food scientist and is of a major concern to the plant manager. The (Freezing) and (Storage) of the products that require low temperatures are the two areas of the production process that contribute to the cost of freezing. These are the LN2 freezer and of course the storage or holding rooms which are using conventional refrigeration. If the storage room's temperature is maintained at − 20°F, the product frozen in the LN2 freezer should not be frozen to an equilibrated temperature less than − 20°F. To minimize the freezing cost in the LN2 freezer, often the product is frozen to a higher temperature such as − 10°F. The product is than allowed to equilibrate to the required

– 20°F temperature in the storage room. The time the product remains on the production floor before it is placed in the holding freezer should be as short as possible. At the most 5 minutes.

> *Do not freeze the product to a temperature lower than the storage room temperature.*
> *This is wasted energy and will contribute to a higher freezing cost.*

14.7 A WORD ABOUT LN2 vs. LCO2

Too often, I am asked to explain the differences between the use of LN2 and LCO2 in food freezers. Although this writing was primarily intended to cover the use of Liquid Nitrogen and freezers using LN2, I feel compelled to add a few words about LCO2 and how it compares to LN2. Carbon Dioxide or CO_2 in its liquid form, LCO2, was after all the first non-conventional refrigerant used by the food processors. LN2 did not come into play until the sixties and very reluctantly at that.

14.7.1 Physical Properties of Carbon Dioxide

Chemical	Carbon Dioxide
Symbol	CO_2
Mol. Weight	44.01
N Boiling point (1 atm.)	- 109.3°F. (- 78.5°C)
Latent Heat Vaporization	245.5 BTU/Lb. (571.3 KJ/Kg.)
Gas phase – SG (32°F & 1 atm)	1.524. (1.539 @ 0°C & 101.325 kPa)
Gas phase – Specific Heat (Cp)	0.199 BTU/Lb °F (0.85 KJ/Kg °C.
Liquid Phase – SG	1.18
Liquid phase – Specific Heat (Cp)	
Triple Point – Temperature	- 69.9°F (- 56.6°C)
Triple Point – Pressure	75.1 psia (517 kPa abs)

CO_2 as a gas is slightly toxic, colorless with a slightly pungent taste. It is a constituent of the air we breathe but only in a minute quantity of 0.038% (380 ppm) The air we exhale contains as much as 4% of CO_2. CO_2 gas combines with water to form a mild acid, called Carbonic Acid

The large quantities used by numerous industries are manufactured by purifying CO_2 rich waste gases from combustion or biological processes. The most common sources are from plants producing hydrogen or ammonia from natural gas, or other hydrocarbon feedstock.

14.7.2 Energy from LN2 & LNCO2

To understand the two different freezing media, let us begin with an illustration of the theoretical energy contents of LN2 and LCO_2.

Cryogenic freezing uses LN2 which at atmospheric pressure has a boiling point of – 320°F (-

196°C) with a heat of vaporization of 85.4 BTU/Lb. In freezer installations, the storage pressure is at the most 20 PSIG (1.4 Bar). If we now warm the gas in the freezer to 0°F before it is exhausted, we get an accumulated energy value of 83.9 + (- 315.6° - 0°) x 0.253 = 163.75 Lb. of LN2.

Liquid Carbon Dioxide (LCO_2) cannot maintain its physical state in the atmosphere (1 atm, 14.7 PSIG) It is injected as a liquid under high pressure of 290 PSIG through nozzles into the tunnel. Upon entering the tunnel, it will form 47% of solid dry ice or CO_2 snow having a temperature of – 109°F, and 53% of gas. The snow has a Latent Heat of Vaporization of 246 BTU/Lb.

Primarily, this snow is used to freeze the product.

Liquid Carbon Dioxide is stored at 290 PSIG and – 4°F, in a receiver, which is similar in construction than the receivers used for the storage of LN2.

The heat of vaporization or heat of sublimation is now 47%x246 = 116 BTU/Lb. Warming the gas to 0°F adds another {-109° - (-70°)} x 0.20 = 8 BTUs, for a total of 124 BTU/LB. NOTE: The colder exhaust temperature of – 70°F.

> NOTE: *The construction of a CO_2 freezer makes it practically impossible to withdraw the spend vapors at the high temperatures as we have seen with the LN2 counter-flow freezers. CO_2 tunnels are configured as isothermal or confection freezers.*

Both the LN2 and the LCO_2 freezers depend on the temperature management of the exhaust vapors on how much energy is derived from one unit of liquid.

Due to the large differences of the heat of vaporization of LN2 (83.9 BTU/LB.) and Liquid/Solid CO_2 (254 BTU/LB.) the tunnels are designed differently. The LN2 tunnel has to "Harness" the huge amount of energy or sensible heat (- 306°F) of the vapors in a "Pre-Cool" zone of a counter flow tunnel. See Fig. 24A. This is the most effective design for a LN2 tunnel.

On the other hand, LCO_2 tunnels are designed to take advantage of the large amount of BTUs available from the heat of sublimation of the solid CO_2 and are designed as convection or isothermal freezers. Spiral freezers are a good example of a convection or isothermal freezer with a set point temperature of about – 80°F for LCO_2.

LCO_2 freezers are not simple to operate, specifically tunnel freezers. The freezer must be sized precisely correct for a specific product and throughput. Any deviation from the original design parameters such as production capability in LBS/Hr. and the product's temperature can cause the freezer to build excessive "snow", which is dry ice, blockages. It requires a more vigilant operator to run a LCO_2 freezer efficiently. Since LCO_2 converts to dry ice when it is released into the atmosphere, very often "Snowing" problems are encountered which can be harmful when it is deposited on the production floor. This deposit of unused dry ice will upset the possibility to a correct prediction of the consumption rate. In general, LCO_2 freezers are less user-friendly, are complicated, and require more maintenance.

LN2 freezers, with the exception of the immersion freezers, are very user-friendly, easy to operate and still able to function efficiently with other than the original design parameters. The set point temperature of – 160°F, can be over ridden for short production runs when production spikes are required without worrisome side effects. Although consultation with the manufacturer is recommended. One negative is the absence of odor of the nitrogen vapors. CO_2 vapors have a distinct pungent odor and are quickly detected when excess vapors escape into the processing room. Generally, both LCO_2 and LN2 freezing equipment should have an oxygen-monitoring device installed in the processing room to avoid mishaps. See chapter 12 for detailed safety instructions.

Similar as with LN2, the LCO_2 supplier should be the one with a plant as close to the processing plant as possible. In addition, make sure they are able to supply you with an uninterrupted supply, during summer and winter months. Do not hesitate to ask for written guaranties for these provisions.

15.   USEFUL INFORMATION

15.0 Remember these Useful Points when,

Calculating the freezer size

Before calculating the freezer size, make sure you know,

A.   The belt loading

B.   The dwell time in minutes (Preferably from tests)

C.   The required maximum production rate

D.   Is the product able to stay on the belt with the maximum circulation?

E.   Are product guide rails required?

F.   Is the product wrapped?

Calculating the LN2 consumption

Before calculating the LN2 consumption, make sure you know,

A.   The tank pressure the "system" is going to use

B.   The BTU's from the LN2 at that pressure (Use Table 2)

C.   The exhaust vapor temperature

D.   The product's energy requirement. (BTU/Lb.)

Calculating the LN2 tank size

Before calculating the tank size, make sure you know,

A.   Distance from air separation plant to the food processing plant

B.   Hours the freezer is in use per operating day

C.   How many pounds or gallons of LN2 is required per day

D.   To get a fill at least once a week

Calculating the LN2 pipe

Before calculating the size (diameter) of the LN2 pipe, make sure you know,

A.   The LN2 requirement to freeze the product in Lbs/hr.

B.   The tank pressure

C.   The *equivalent* length of pipe between the tank and the freezer

D.   When using Polyurethane insulated piping, make sure the pipe insulation is at least 4" thick. (See Section 8, Chapter 8.4.0)

E.   Use the very minimum of elbows and tees

F.   Use full-port ball valves with extended stem only.

About 30 years of experience have contributed to the content of this manual. Sadly to say, that during these years, I have observed that the majority of freezers are bought without any knowledge of cryogenic principles. When it is cheap, anything resembling a freezer will do. Do not make this mistake! Consult with qualified engineers (More than one is advisable) before you make a decision. Eventually, the freezer will be a pivotal part of your process, determines your bottom line, and hence, profits. Profits can only be realized when the correct freezer and system have been selected.

15.1 Conversion Tables

Pressure

1 pound/sq.in	= 6890 Pa	= 6.895	kPa.
1 pound/sq.in	= 0.0689		bar
1 bar	= 14.504		psi
1 pound/sq.in	= 27.67		in w.g.
1 atm	= 101.325		kPa
1 atm	= 14.696		lb/sq in.
1 ft. water	= 0.4335		lb/sq in.
1 lb/sq in.	= 2.31		ft. water

Heat (Energy) quantity

1 BTU raises the temperature of 1 pound of water 1 degree Fahrenheit.

1 BTU	= 1,055 joules	= 0.252	kcal
1 kcal	= 3.97 BTU		
1 HP-hour	= 2,544.5 BTU		

kcal/kg x 1.8 = BTU/lb.

kjoules/kg x .43 = BTU/lb.

Specific Heat: BTU/lb. °F (1 Btu/lb °F = 1 kcal/kg.°C)

Conductivity: BTU/ft.hr.°F (1 BTU/ft.hr.°F = 1488 kcal/h.m.°C.)

Convection: BTU/sq.ft.hr.°F (1 BTU/sq.ft.hr.°F = 4.887 kcal/h.m2.°C.

Temperature

1°F = 5/9°C   32°F = 0°C   °F = (1.8 x °C) + 32

Mechanical Refrigeration data

1 kw provides:

13,900 BTU/hr. at 32°F

10,000 BTU/hr. at 14°F

6,800 BTU/hr. at -4°F

5,200 BTU/hr. at -22°F

4,000 BTU/hr. at -40°F

TEMPERATURE CONVERSIONS

(-459.4 to 0)			(1 to 60)			(61 to 290)			(300 to 890)			(900 to 3000)		
C	F/C	F	C	F/C	F	C	F/C	F	C	F/C	F	C	F/C	F
-273	-459.4		-17.2	1	33.8	16.1	61	141.8	149	320	572	482	900	1652
-268	-450		-16.7	2	35.6	16.7	62	143.5	154	310	590	488	910	1670
262	-440		-16.1	3	37.4	17.2	63	145.4	160	320	608	493	920	1688
257	-430		-15.6	4	39.2	17.8	64	147.2	166	330	626	499	930	1706
251	-420		-15.0	5	41.0	18.3	65	149	171	340	644	504	940	1724
-246	-410		-14.4	6	42.8	18.9	66	150.8	177	350	662	510	950	1742
-240	-400		-13.9	7	44.6	19.4	67	152.6	182	360	680	516	960	1760
-234	-390		-13.3	8	46.4	20	68	154.4	188	370	698	521	970	1778
-229	-380		-12.8	9	48.2	20.6	69	156.2	193	380	716	527	980	1796
-223	-370		-12.2	10	50.0	21.1	70	158	199	390	734	532	990	1814
-218	-360		-11.7	11	51.8	21.7	71	159.8	204	400	752	538	1000	1832
-212	-350		-11.1	12	53.6	22.2	72	161.6	210	410	770	549	1020	1868
-207	-340		-10.6	13	55.4	22.8	73	163.4	216	420	788	560	1040	1904
-201	-330		-10.0	14	57.2	23.3	74	165.2	221	430	806	571	1060	1940
-196	-320		-9.4	15	59.0	23.9	75	167	227	440	824	582	1080	1976
-190	-310		-8.9	16	60.8	24.4	76	166.8	232	450	842	593	1100	2012
-184	-300		-8.3	17	62.6	25	77	170.6	238	460	860	604	1120	2048
-179	-290		-7.8	18	64.4	25.6	78	172.4	243	470	878	616	1140	2084
-173	-280		-7.2	19	66.2	26.1	79	174.2	249	480	896	627	1160	2120
-169	-273	-459.4	-6.7	20	68.0	26.7	80	176	254	490	914	638	1180	2156
-168	-270	-454	-6.1	21	69.8	27.2	81	177.8	260	500	932	649	1200	2192
-162	-260	-436	-5.6	22	71.6	27.8	82	179.6	266	510	950	660	1220	2228
-157	-250	-418	-5	23	73.4	28.3	83	181.4	271	520	968	671	1240	2264
-151	-240	-400	-4.4	24	75.2	28.9	84	183.2	277	530	986	682	1260	2300
-146	-230	-382	-3.9	25	77.0	29.4	85	185	282	540	1004	693	1280	2336
-140	-220	-364	-3.2	26	78.8	30	86	186.8	288	550	1022	704	1300	2372
-134	-210	-346	-2.8	27	80.6	30.6	87	188.6	293	560	1040	732	1350	2462
-129	-200	-328	-2.2	28	82.4	31.1	88	190.4	299	570	1058	760	1400	2552
-123	-190	-310	-1.7	29	84.2	31.7	89	192.2	304	580	1076	788	1450	2642
-118	-180	-292	-1.1	30	86.0	32.2	90	194	310	590	1094	816	1500	2732
-112	-170	-274	-0.6	31	87.8	32.8	91	195.8	316	600	1112	843	1550	2822
-107	-160	-256	0.0	32	89.6	33.3	92	197.6	321	610	1130	871	1600	2912
-101	-150	-256	0.6	33	91.4	33.9	93	199.4	327	620	1148	899	1650	3002
-96	-140	-238	1.1	34	93.2	34.4	94	201.2	332	630	1166	927	1700	3092
-90	-130	-220	1.7	35	95.0	35	95	203	338	640	1184	954	1780	3182
-84	-120	-184	2.2	36	96.8	35.6	96	204.8	343	650	1202	982	1800	3272
-79	-110	-166	2.8	37	98.6	36.1	97	206.6	349	660	1220	1010	1850	3362
-73	-100	-148	3.3	38	100.4	36.7	98	208.4	354	670	1238	1038	1900	3452
-68	-90	-130	3.9	39	102.2	37.2	99	210.2	360	680	1256	1066	1950	3542
-62	-80	-112	4.4	40	104.0	37.8	100	212	366	690	1274	1093	2000	3632
-57	-70	-94	5.0	41	105.8	43	110	230	371	700	1292	1121	2050	3722
-51	-60	-76	5.6	42	107.6	49	120	248	377	710	1310	1149	2100	3812
-46	-50	-58	6.1	43	109.4	54	130	266	382	720	1328	1177	2150	3902
-40	-40	-40	6.7	44	111.2	60	140	284	388	730	1346	1204	2200	3992
-34	-30	-22	7.2	45	113.0	66	150	302	393	740	1364	1232	2250	4082
-29	-20	-4	7.8	46	114.8	71	160	320	399	750	1382	1260	2300	4172
-23	-10	14	8.3	47	116.6	77	170	338	404	760	1400	1288	2350	4262
-17.8	0	32	8.9	48	118.4	82	180	356	410	770	1418	1316	2400	4352
			9.4	49	120.2	88	190	374	416	780	1436	1343	2450	4442
			10.0	50	122.0	93	200	392	421	790	1454	1371	2500	4532
			10.6	51	123.8	99	210	410	427	800	1472	1399	2550	4622
			11.1	52	125.6	100	212	413.6	432	810	1490	1417	2600	4712
			11.7	53	127.4	104	220	428	438	820	1508	1454	2650	4802
			12.2	54	129.2	110	230	446	443	830	1526	1482	2700	4892
			12.8	55	131.0	116	240	464	449	840	1544	1510	2750	4982
			13.3	56	132.8	121	250	482	454	850	1562	1538	2800	5072
			13.9	57	134.6	127	260	500	460	860	1580	1566	2850	5162
			14.4	58	136.4	132	270	518	466	870	1598	1593	2900	5252
			15.0	59	138.2	138	280	536	471	880	1616	1621	2950	5342
			15.6	60	140.0	143	290	554	477	890	1634	1649	3000	5432

Deg.F=1.8 (Deg.C) + 32   Deg.F= 9/5K - 459.67   Deg.K= 5/9F + 459.67

15.3 REFRIGERATION CAPACITY OF LIQUID NITROGEN (Calculated)
(Q)

Exhaust Temp.		Tank Pressure							
		Atm. Press. 1.013 bar 14.7 psig		1 bar rel. 14.5 psig		1.5 bar rel. 21.7 psig		2 bar rel. 29 psig	
C	F	kcal/l.	BTU/lb.	kcal/l.	BTU/lb.	kcal/l.	BTU/lb.	kcal/l.	BTU/lb.
-60	-76	65.9	147	67	149	62.3	139	61	136
-50	-55	68.2	152	69	154	64.6	144	64	142
-40	-50	69	154	70	156	65	145	64	143
-30	-20	72	161	73	163	69	153	67.3	150
-20	-10	73.5	164	74.4	166	70	155	69	153

1 Kcal /l. = 2.23 BTU/lb.

Average room temperature: 80°F, 27°C, 300 K

Water boils at: 212°F, 100°C, 313 K

Water freezes at: 32°F, 0°C, 273 K

16. DEFINITIONS

1.   Enthalpy

2.   Specific Heat or Heat Capacity

3.   Latent Heat of Fusion

4.   Latent Heat of Vaporization

5.   Normal Boiling Point

6.   Sublimation

7.   Flash Off

8.   Thermal Conductivity

9.   Heat Transfer

10. "U" Factor

1. **ENTHALPY**

    Enthalpy is a measure of heat in the system. Scientists figure out the mass of a substance when it is under a constant pressure. Once they figure out the mass, they measure the internal energy of the system. All together, that energy is the enthalpy. They use the formula, $H = U + PV$.

    Where H is the enthalpy value, U is the amount of internal energy, and P and V are pressure and Volume of the system.

2. **SPECIFIC HEAT or HEAT CAPACITY**

    C – The amount of heat necessary to raise (Lower) the temperature of a unit mass of material by one degree. Expressed in $J/Kg/°K$ or $BTU/lb/°F$.

    The Specific Heat of water is $1\ BTU/lb/°F = 1\ calorie/gram\ °C = 4.186\ joule/gram\ °C$ which is higher than any other common substance.

    Without a phase change the relationship between heat and temperature change is usually expressed in the form shown below.

    $Q = c.m.\Delta T$

    Heat added = specific heat x mass x $(t_{final} - t_{initial})$

3. **LATENT HEAT OF FUSION**

    The amount of heat released or absorbed at a specific temperature when the unit mass of a material is transformed from one state to another.

    Expressed in joule/kg or BTU/lb.

    Transforming water to ice requires 144 BTU/lb. (L = 144 BTU/lb.)

4. **LATENT HEAT OF VAPORIZATION**

    The quantity of heat required to convert a unit of mass of a material from the liquid to the gaseous state at a given pressure and temperature.

5. **NORMAL BOILING POINT**

    The temperature at which a liquid boils under a pressure of one atmosphere or 14.7 psig. LN2 = -320.36°F. (-195.76°C) CO_2 = -109.25°F. (-78.47°C)

6. **SUBLIMATION**

    The direct passage of a substance from the solid state to the gaseous state without any time appearing in the liquid state. Such as with CO_2 at atmospheric pressure.

7. **FLASH OFF**

The transformation of a liquid at a certain pressure into a gaseous state when released to a lower pressure. Less pressure differential will result in less flash off.

8. **THERMAL CONDUCTIVITY**

The rate at which heat or cold flows through a material by conduction and is expressed as
$Q = K.A (\delta T/\delta X)$
Where Q is heat flow (J/s), A is the area of heat transfer normal to the heat flow, $(\delta T/\delta X)$ is the temperature gradient in the X direction, and the proportionality constant, K, is the thermal conductivity (W/m °K)

9. **HEAT TRANSFER**

Heat flows from a warm to a colder area by three different methods:

1. By conduction from molecule to molecule within a substance or between touching substances

2. By radiation or heat (cold) waves

3. By convection or eddy currents in or movements of liquids and gases.

In practical calculations, heat transfer is usually obtained by two or all three methods.

10. **"U" FACTOR**

For practical purposes, engineers often lump the above factors into an over-all "Coefficient of heat transfer" called the U factor. Expressed in BTU/hr/ft²/°F.
Every cryogenic freezer has a given U factor. (Best obtainable from the manufacturer.)

11. **Energy**

The ability to do work or produce heat.

12. **A "Smart" operator**

An individual with the ability to decide that the best his company is served is to stay away from freezing equipment older than 2 years and never but never attempts to build his own freezer.

17. LN2 FREEZING SYSTEM RECORDS

17.1　The LN2 Receiver or Tank

Date Installed: ....

Model number: ....

Natl Board Number: ....

Date Manufactured: ....

Vacuum readings:

Warm Tank: ....

Cold Tank: ....

Safety Valve Ratings:

Main Blow off: ....

Weep Settings: ....

Capacity in:

Gallons: ....

Pounds: ....

Standard Cubic Feet: ....

Lowest possible Tank Pressure: ....

Level Gauge:

Full at: ....

Reorder at: ....

17.2 The Piping System

Date of Installation: ....

Installation Contractor: ....

Total equivalent length: ....

Vacuum or PU Insulated: ....

Nominal Pipe size: ....

Safety Valve Rating: ....

17.3 The Freezer

Date Installed: ....

Make: ....

Serial Number: ....

Model Number: ....

Date Manufactured: ....

Number of Nozzles: ....

Manufacturer: ....

Size: ....

Model Number: ....

Type: ....

Capacity at Tank Pressure: ....

Control Valve:

Make: ....

Nominal Full Port Size: ....

Capacity at 20 PSIG: ....

Cv at 20 PSIG: ....

NOTE: 20 PSIG is the assumed tank pressure

17.4   The Exhaust System

Date Installed...:

Make:..

HP: ....

Voltage:...

Capacity in SCF/Min:..

Inlet Opening: ....

Outlet Opening: ....

Duct Size to Freezer:...

17.5   Recommended Settings:

Tank Pressure Control Panel:...PSIG

Date set:..

Date set:..

Set Point temperature:

Product A:...°F

Product B:...°F

Product C: ....°F

17.6   Important Contacts

LN2 reorder number and person:...

Tank maintenance contact:..

Piping installation contact:..

Freezer contact: ....

Consultant: ....

AUTHOR BIOGRAPHICAL NOTE

Cornelis J.F.Elenbaas was born in 1940 and immigrated to the United States from Holland in 1997. He grew up in Zwolle, Holland where he finished high school and finished his education from the Dutch Merchant Marine Academy with a major in mechanical engineering.

After a seven-year period in the Dutch Merchant Marine, he immigrated to Chicago, IL. where he worked for a firm involved in the manufacture and marketing of industrial gases. A five-year transfer to Europe as the international manager for cryogenic applications broadened his knowledge of cryogenics and the global importance of that technology for many industries. Following his return to the US, he has held the positions of Marketing Manager Cryogenic applications for North America and the last 13 years as Engineering Director for Cryogenic applications. He is now retired and is finally enjoying his three daughters and five grand children.